IMAGES
of America

ITAWAMBA COUNTY

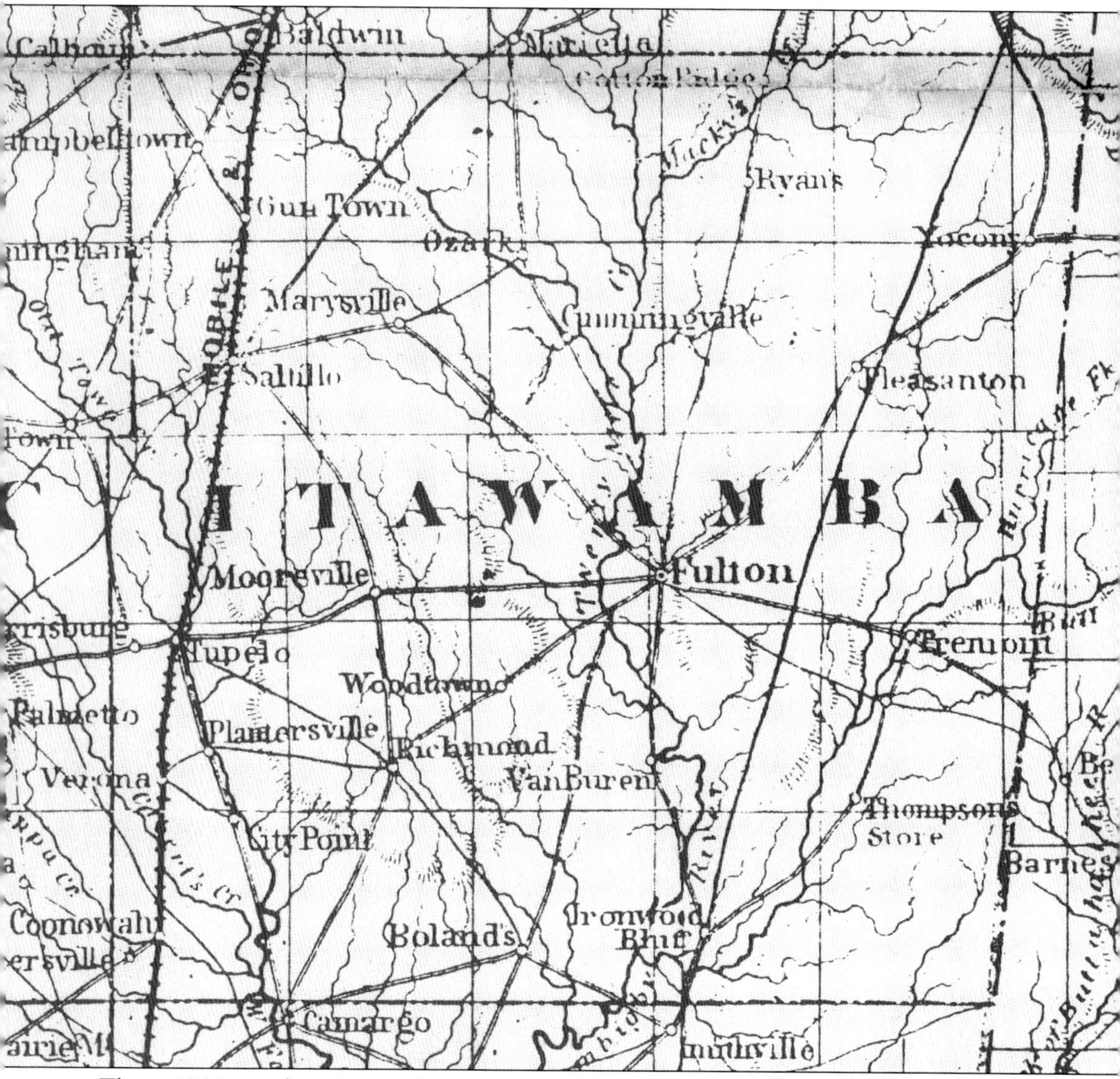

This c. 1866 map shows Itawamba County's original borders. Westward from the Alabama state line, the county formerly included Harrisburg and neighboring Tupelo, near the border with Pontotoc County.

On the Cover: From 1903 until 1905, Walter Clifton Cummings Sr. owned and operated this sawmill west of the village of Mantachie, in Mantachie Swamp. In an obviously posed photograph, mill workers take a break from the hard labor and inject humor into their poses, while logs at the rear of the mill are waiting to be turned into lumber. (Courtesy of Coleen Thornton, great-granddaughter of W.C. Cummings.)

IMAGES
of America

ITAWAMBA COUNTY

Mona Robinson Mills

ARCADIA
PUBLISHING

ISBN 978-0-7385-9068-4

Published by Arcadia Publishing
Charleston, South Carolina

Printed in the United States of America

Library of Congress Control Number: 2012945084

For all general information, please contact Arcadia Publishing:
Telephone 843-853-2070
Fax 843-853-0044
E-mail sales@arcadiapublishing.com
For customer service and orders:
Toll-Free 1-888-313-2665

Visit us on the Internet at www.arcadiapublishing.com

This book is dedicated to my children—
Alysson, Chip, Rebekah, and Penn.
Your ancestors are surely proud;
their labor and toil has not been for naught.

Contents

Acknowledgments

Making new friends, meeting with new kinfolk as well as familiar ones, and collecting old photographs and historical information from the people of Itawamba County has been a pure labor of love. The project could not have happened without the assistance of the Itawamba County Historical Society, which provided many of the photographs included in the book, and especially the society's librarian, Marilyn Leary, whose patience and infectious excitement were very much appreciated. Because of the society's assistance with this book and due to the role the society has played in preserving our county's history, a portion of the royalties from the sale of this book will go to the Itawamba County Historical Society.

Many thanks go to Rubye Del Harden, who generously allowed me to comb through the enormous collection of photographic negatives accumulated by her late father, Delmus Harden, the owner and former publisher of the "only newspaper in the world that cares anything about Itawamba County," the *Itawamba County Times*. These negatives provide a remarkable look at Itawamba County through Delmus's eyes into the last half of the 20th century.

So many people searched through closets and drawers, and asked cousins, grandparents, and others about old photographs, that I cannot thank them all by name, although Virble and Charles Booth, the "first couple" of the Itawamba County Historical Society, deserve recognition not only for their help with this project but also for all they have done, and still do, to preserve the history of our county.

Special thanks are extended to my father's first cousin, Lucy Robinson, who started my love affair with old photographs when she graciously gifted me with the photographic contents of her grandmother's trunk—wonderful treasures to be shared and smiled over, time and time again.

And finally, I thank my husband, whose encouragement and support of me never waivers, regardless of whatever project or task I undertake, and whose own love for and knowledge of our home county inspires me.

Unless otherwise noted, all images come from the Itawamba County Historical Society or the author's personal collection.

INTRODUCTION

The lands that comprise what is now north Mississippi were purchased from the Chickasaw Indian Nation in 1832 under the terms of the Treaty of Pontotoc, in which the Chickasaws agreed to sell to the United States all of their land east of the Mississippi River. Twelve counties were formed out of this purchase, one of which was named Itawamba, after a Chickasaw leader.

Legend has it that Chief Itawamba earned his name as a young man when he defended his village against a raid from other tribes while the older warriors were gone. For this he was given the honor of sitting on a bench, as opposed to on the ground, at the next tribal council meeting, thus earning the name Itte-wamba Mingo, which means "Bench Chief." In the white community, Chief Itawamba was otherwise known as Levi Colbert, the son of an Indian woman and her Scotsman husband, who was a merchant and large landowner. Levi, as a chief and leader of the Chickasaw tribe at the time, was instrumental in the negotiation of the treaty with the federal government.

Following the treaty, the US government surveyed and divided the land into sections for eventual public sale. News of the treaty reached the people of Georgia, the Carolinas, eastern Tennessee, Virginia, and other states, and many potential settlers moved into western Alabama anticipating the opening of the federal land office in Pontotoc. Finally, in 1836, the Mississippi legislature formed counties out of the former Indian lands, and the land sale began. Some settlers, however, did not wait for official sanction before moving; surveyors' field notes dated 1833 include references to the preexisting homes or farms of John Dulaney, Benjamin Wise, Eliba Allen, and Edward Maxcy in what is now Itawamba County.

Josiah Hinds arrived with his family in 1840, settling on the western side of the Tombigbee River. In his diary, he writes, "We are among strangers in a strange land, and in a wilderness, where but a short time since was heard the yell of the savage, and where the hoot of the owl and the prowl of the wolf is still heard."

At the time of its formation, Itawamba County's borders included the present-day towns and communities of Richmond, Mooreville (originally Hatter Shop), Verona, Saltillo, Shannon (City Point), and Guntown in present-day Lee County. The now extinct village of Harrisburg was also within old Itawamba's boundaries, but its residents and businesses moved one and a half miles east to meet the new railroad, to an area that later became Tupelo. As towns in then western Itawamba County grew, their citizens grew weary of the long trek to the courthouse in Fulton to transact official business, a journey that included a ferry ride across the Tombigbee River or a detour north around the headwaters of the Tombigbee. The county's borders remained unchanged until 1866 when a 10-mile-wide swath of the western boundary was sliced off, along with a portion of eastern Pontotoc County, to form the new county of Lee. Later, in 1870, a three-mile strip of land along the northern boundary was taken away and given to Tishomingo and Prentiss Counties.

But, before there was a county, there was the river. The Tombigbee River created a natural barrier from the earliest days of the county, both physical and mental. Flooding was a constant source of irritation and worry. Crossing the river was not easy; there were ferries, but no bridge existed until 1907, when an iron span was built across the river at Fulton. In March 1844, pioneer Josiah Hinds wrote in his diary, "We have had three days of constant rain, and all the branches and creeks are overflown with water. In going to Fulton yesterday to attend our temperance meeting, I had to leave my horse on this side of Bigbey [Tombigbee] and borrow a canoe, and I then proceeded to cross the river and swamp, which I found to be about one and a half miles across."

The Tombigbee River was the county's only source of transportation, there being no roads, only Indian paths, in the early years. The first villages of the new county were along the river: Wheeling, Ironwood Bluff, and Van Buren. Located on a high bluff along the west bank of the Tombigbee, Van Buren was the largest town in Itawamba County in 1840, boasting a busy river port, boat landing, several stores, and even a town cemetery. However, after the Mobile & Ohio Railroad was built several miles west of Van Buren in the late 1850s, river traffic diminished in favor of the railroad, and the little village of Van Buren declined with it. Much of the county's future population growth took place along the railroad in western Itawamba County.

The earliest affairs of the new county were carried out in the private homes and stores of Van Buren until Fulton, as the county seat, was incorporated on May 11, 1837, by act of the Mississippi legislature. Named for Robert Fulton, inventor of the first commercially successful steamboat, the village was located about one mile east of the Tombigbee River on a bluff overlooking rich bottomlands, on a site that was covered by a thick growth of large pine trees.

Because the river at Fulton was shallow and only marginally navigable during most of the year, Fulton never really grew as a major river town. However, the town prospered as the seat of county government. Fulton also was a stopover on a stagecoach route between the Eastport-Iuka port on the Tennessee River and Cotton Gin Port, a busy shipping center on the Tombigbee River in Monroe County. The first courthouse, a log structure, was erected sometime between 1838 and 1843; by 1850, Fulton had two private schools, two hotels, a newspaper, five attorneys, and four physicians, along with several other businesses.

In his "Olden Times" columns for the *Tupelo Journal* in 1905, W.L. Clayton recalls "Old Fulton" in the pre–Civil War days:

> The social intercourse of the people of that old town in the long ago often took the form of barbeque picnics. The river bottom was near to the town, near enough for all to walk, was well stocked with squirrels and some turkeys, and the river and Cumming's Creek were well supplied with fish, and the citizens would turn out in masse, the women, boys and girls with fishing tackle, and the men with their guns. . . . I remember one of these picnic barbeque hunts . . . We had one hundred and eighty-one squirrels barbequed that day, and they, with the other viands we that day enjoyed, were enough to have satisfied the appetite of a king.

Although Itawambians no longer depend on agriculture for jobs, the county is still quite rural and populated with dozens of small communities. As many as 90 different post offices have existed during the county's history. Places like Boland's, Elfin, Guinea, Miston, and others have faded away from local consciousness. Today, the county has three incorporated towns—Fulton, Mantachie, and Tremont.

The town of Mantachie, in northwestern Itawamba County, grew from a crossroads where Timothy Woods Pearce operated a store a short distance from the Aberdeen-Jacinto Road. Near the store and future town was the site of one of the largest Indian villages in north Mississippi. A nearby creek was called Man-at-chee in honor of a Chickasaw chief, and the town of Mantachie took its name from this creek.

In eastern Itawamba County, Dilmus J. Stone purchased land in 1845, built a log cabin, and established a general store, which he called Stone's Crossroads, the store being located where two main roads—Fulton-Bexar Road and Russellville-Cotton Gin Road—crossed. Later, the crossroads became known as Tremont, although the origin of that name is a mystery.

From the rolling land west of the Tombigbee River to the rugged terrain of the hills hugging the Alabama state line, Itawamba County is blessed with a diverse landscape and kindhearted people. Itawambians have a treasured past in their pioneer ancestors—strong, hardy settlers who endured many hardships to make a life out of the wilderness. These ancestral families produced such familiar names as Elvis Presley, Tammy Wynette, and Jimmie Lunceford, but they also produced sawmillers, churn-makers, farmers, and other strong, independent people who have been proud to call Itawamba home.

One

Scenes from Days Gone By

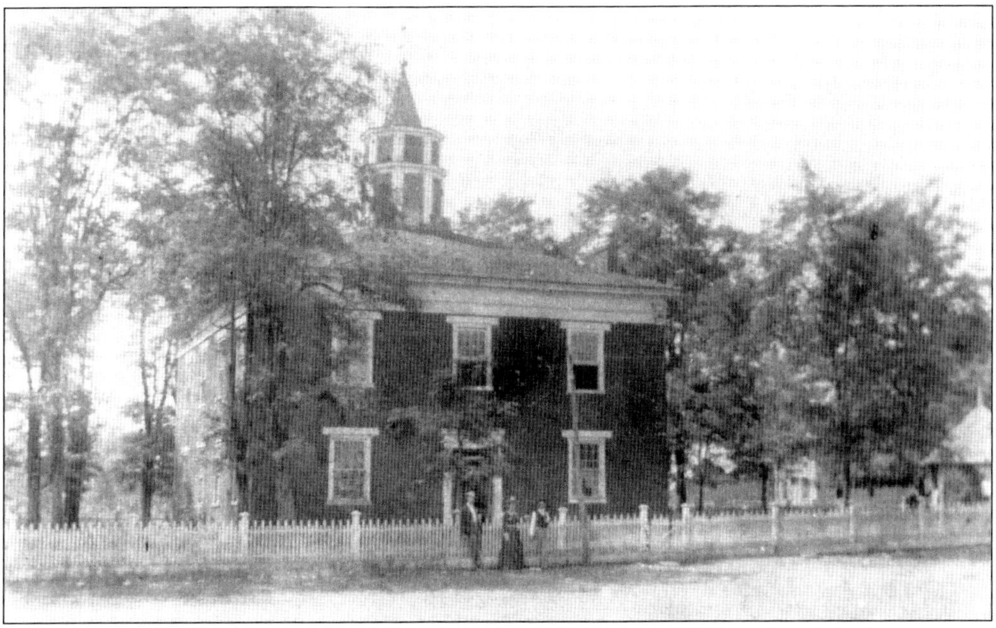

This two-story, Federal-style redbrick building, shown here about 1910, replaced the log structure that was Itawamba County's first courthouse. An octagonal cupola with a weather vane once topped the courthouse. Three chimneys stood on the east and west sides, each with fireplaces on both floors. Heavy double doors opened to a central hall that ran north to south while large windows adorned each exterior wall. The courthouse was surrounded by a wooden fence, with gates centered on all four sides.

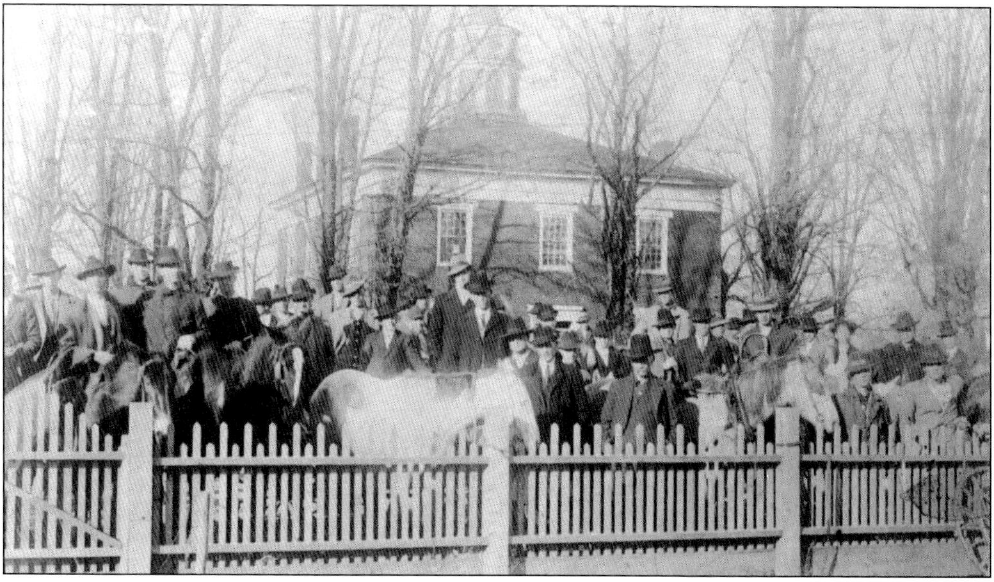

In a wintry scene from about 1920, men surround the south lawn of the courthouse, probably during "court days," when they came to swap horses and tales. Likely built by J.J. Blythe about 1854, the courthouse was similar to the one at Jacinto in old Tishomingo County. Black locust trees were planted around 1875 and completely surrounded the courthouse until subsequent remodels took them down. As late as 1936, seventeen locust trees still stood, many with four-foot circumferences.

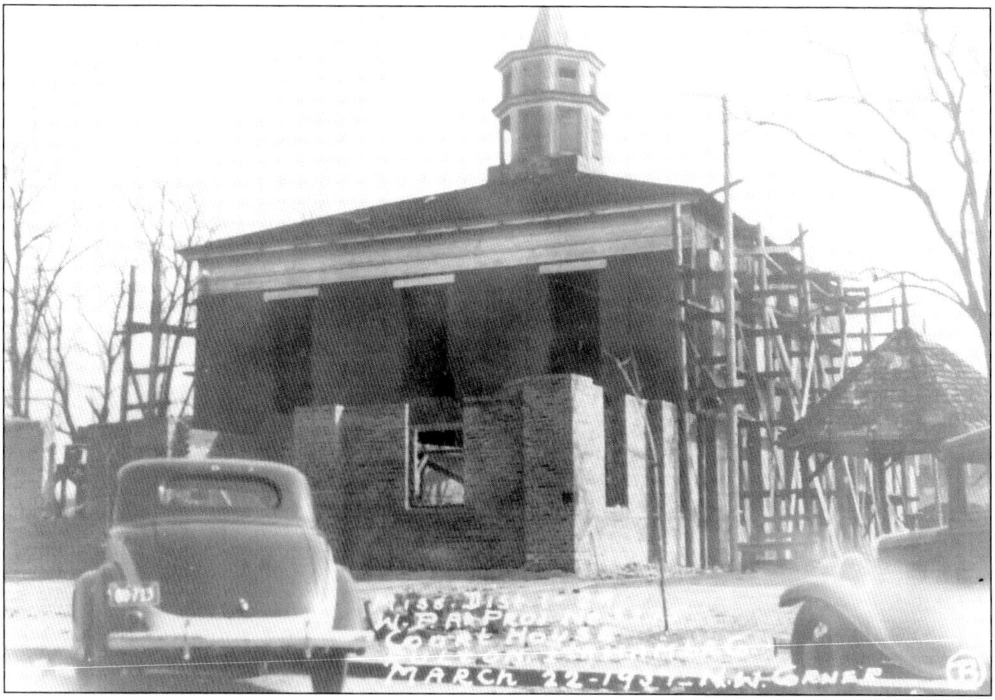

In 1936, an extensive remodeling and enlargement of the courthouse was undertaken as a project of the Works Progress Administration. During the renovation, court was held in the high school auditorium. Along with adding 42 feet to the south side of the building, a hall was built running east and west through the middle of the courthouse. The shed on the right housed a public well.

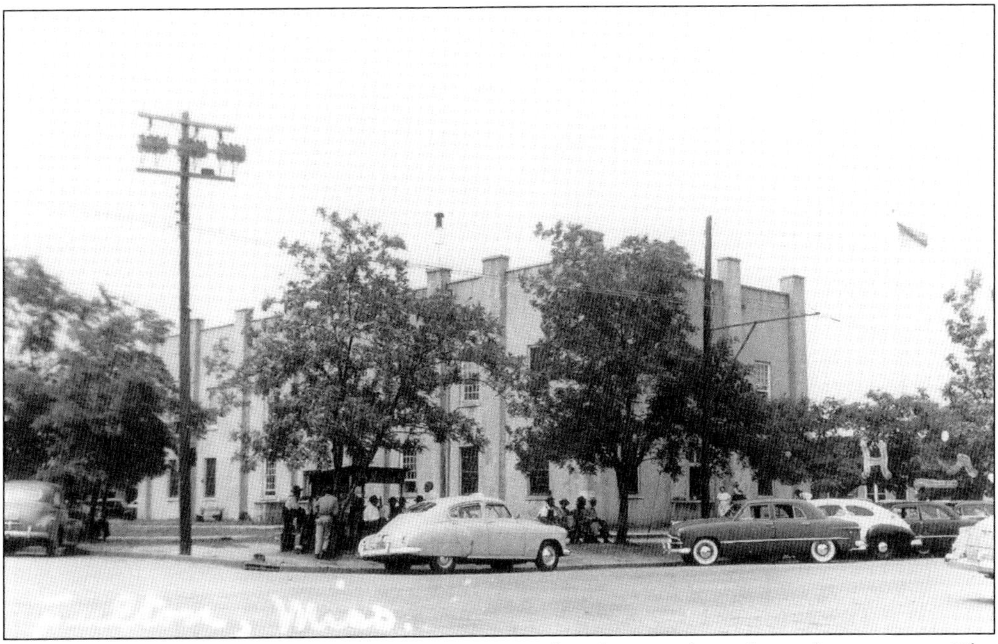

Dated 1949, a postcard photograph shows a taxi stand at the courthouse's northeast corner. The trees in this photograph were likely from a 1933 beautification project in which young locust trees were planted next to the curb; the older locust trees appear to have been removed. Notice the exterior of the courthouse now has a stucco treatment. (Courtesy of Julian Riley.)

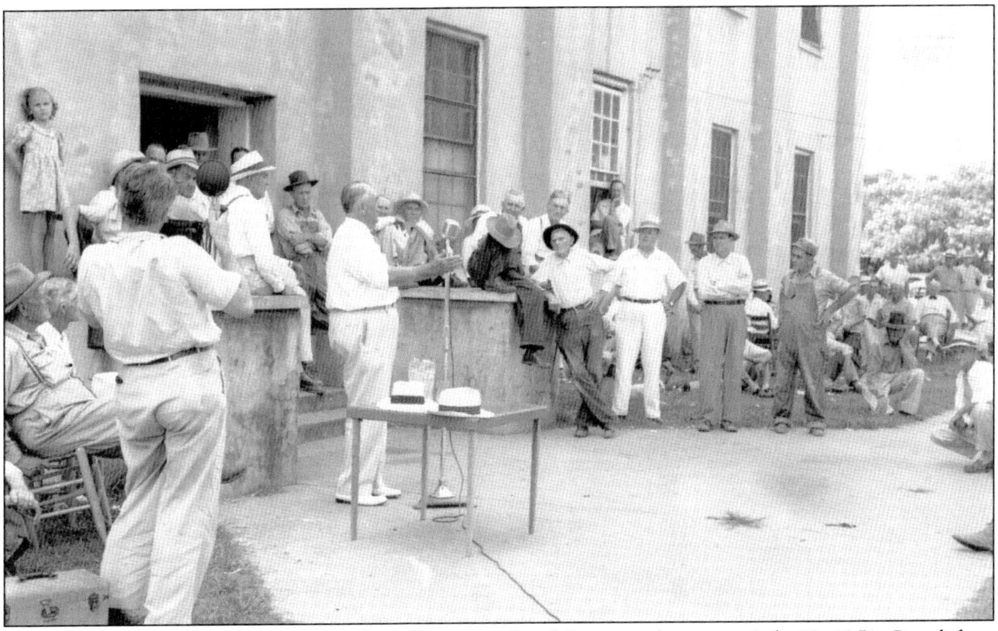

Gov. Hugh White speaks to a crowd on the east side of the courthouse on July 12, 1951. Candidate speakings (speechmaking) usually drew huge crowds wherever they were held. Note the peeling paint on the courthouse wall; records show that another courthouse remodeling project was undertaken in 1952 to add a two-story annex on the west side of the building at a cost of $12,461. (Courtesy of the Delmus Harden family.)

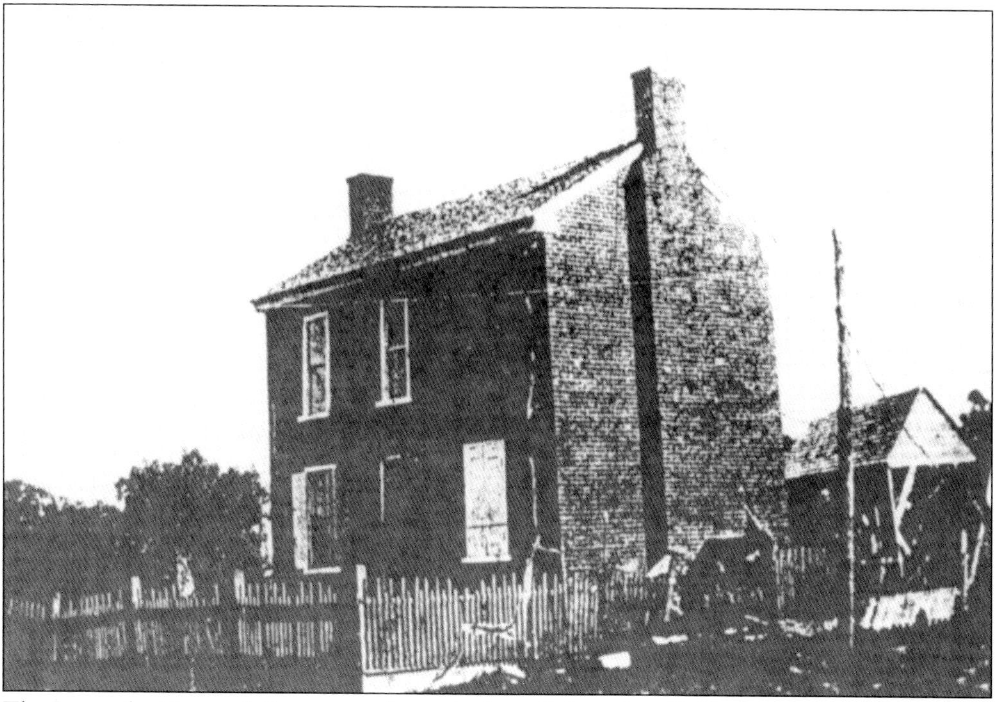

The Itawamba County Jail was once located where Fulton City Hall sits today, at the corner of Cummings and Wiygul Streets. The two-story building was supposedly constructed at the same time as the courthouse, around 1854, using brick fired from local clay. This brick structure served as the county jail for nearly 100 years until burning down in 1942. A new facility was erected just east of the old jail.

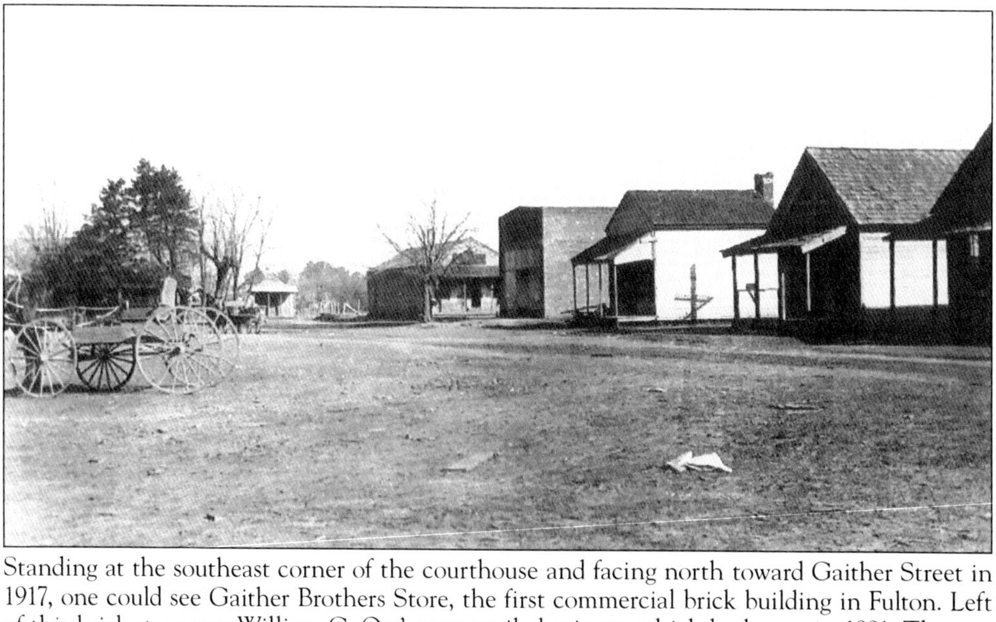

Standing at the southeast corner of the courthouse and facing north toward Gaither Street in 1917, one could see Gaither Brothers Store, the first commercial brick building in Fulton. Left of this brick store was William G. Orr's mercantile business, which he began in 1891. The two buildings to the right later housed sandwich shops, while the small building in the rear of the photograph was Green's Coffin Shop.

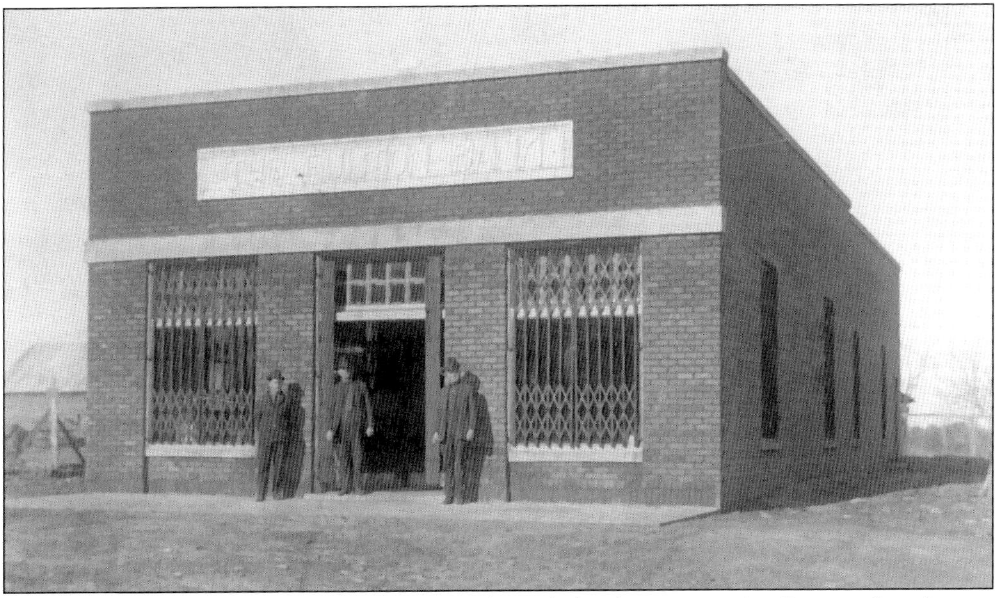

When established in 1905, Fulton Bank was the only bank in Itawamba County. Standing in front of their new steam-heated, brick building in 1921 are J.C. Whitehead Sr., Prentiss Howard, and Richard Senter. Whitehead was president of the Fulton Bank when it merged with Bank of Tupelo. His son, Itawamba-born J.C. Whitehead Jr., became president of the Bank of Tupelo in 1959 and led the bank as it expanded and became Bank of Mississippi. Below, customers wait to conduct business at the tellers' windows of Fulton Bank. The hatless man with the vest is Manuel Gorden, a Russian Jew who emigrated following World War I and opened Gorden's Department Store on the south side of the courthouse square. Others in the photograph are unidentified. (Below, courtesy of the Delmus Harden family.)

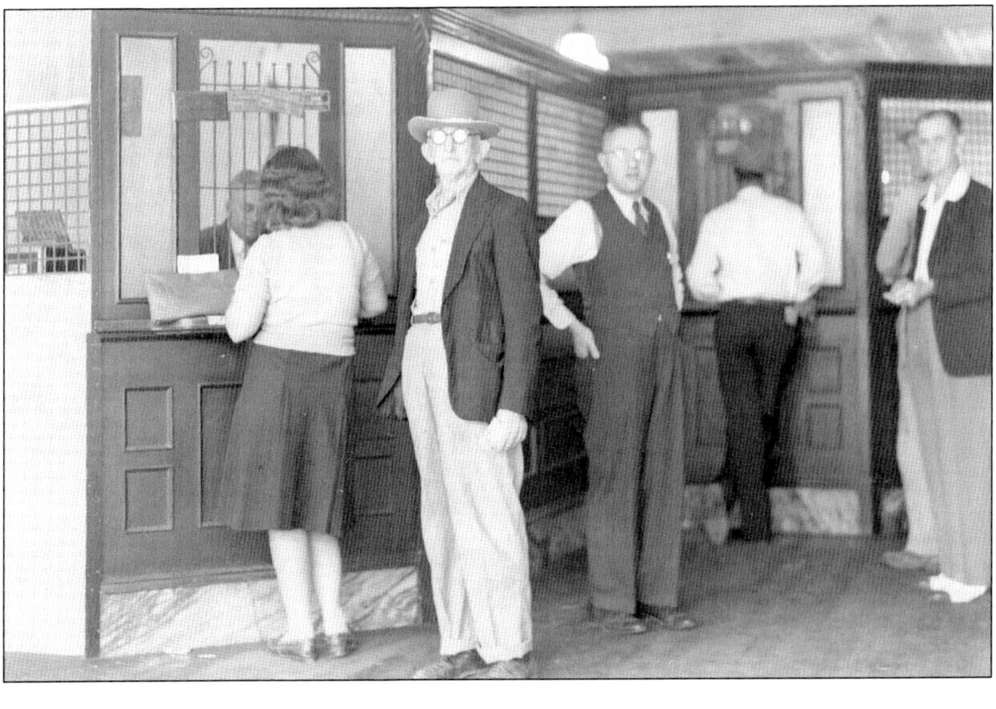

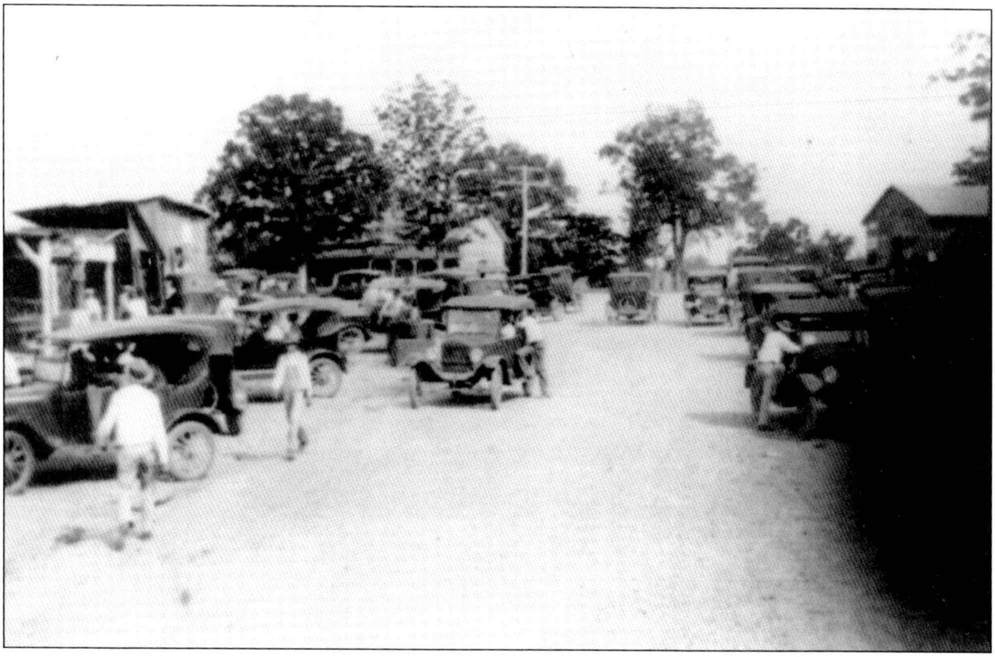

In these street scenes from the 1920s, Mantachie is a bustling place. Note the number of cars and people conducting business. In the background of the above photograph, one can see the Pearce-McFerrin home and a nearby telephone pole. Below, people gather on the porch of the two-story, wood-frame building that once housed the local Woodmen of the World organization. To the left of the two-story building is a grist mill. (Both, courtesy of the Arthur Kermit McFerrin family.)

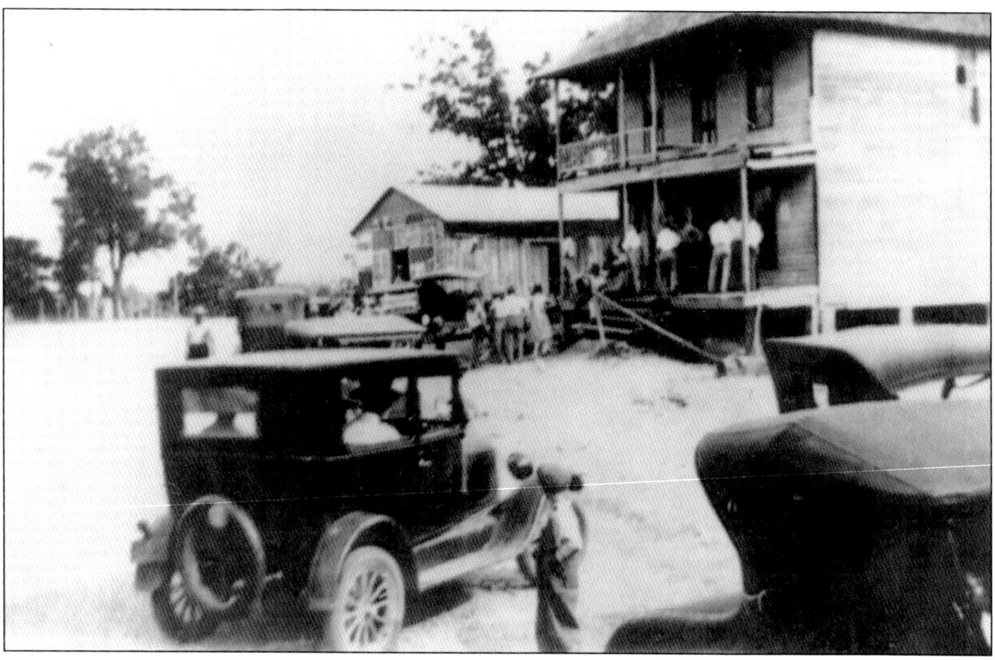

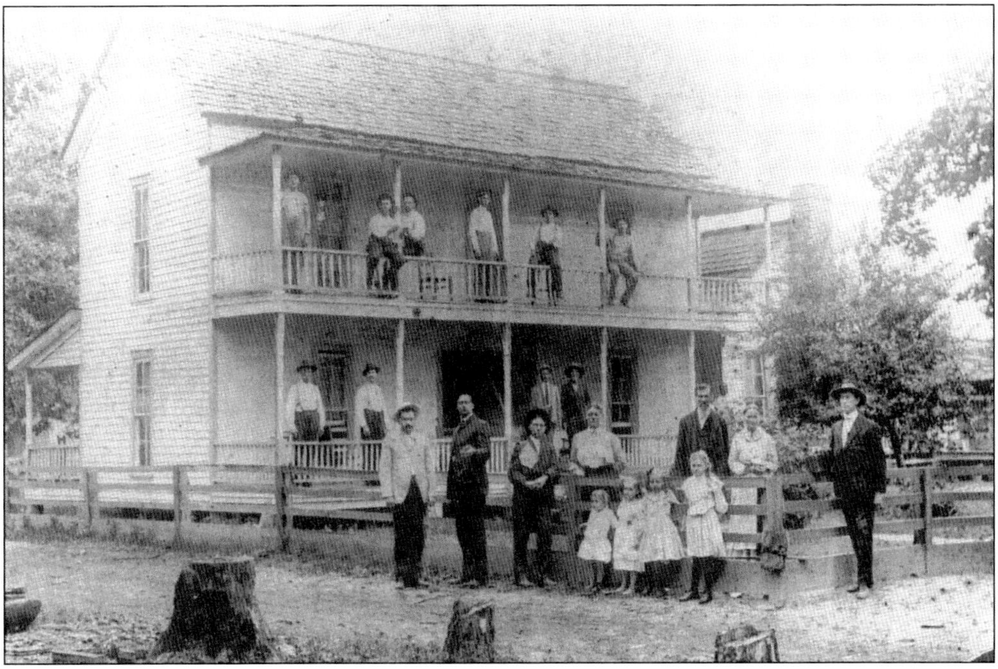

Mantachie was originally settled near Center Star Cemetery, but the town moved southeast to its present location around 1885. The Mantachie Hotel, pictured here around the turn of the 20th century, was owned by Ben East in the 1910s and 1920s. The hotel enjoyed a brisk business from salesmen who, because of poor roads, remained overnight in town. Pictured are, from left to right, (standing in front) J.E. Sims, Mr. McMillen (photographer), Dr. Noonan Nanney, Bell Austin Morgan and her four daughters, Mr. and Mrs. Willis Austin, and Dr. Ford; (on the first-story porch) Billy Waters, T.E. Jones, an unidentified salesman, and Timothy W. Pearce; (on the second-story porch) Bill Ford, Tom Gregory, Lee Waters, Sam Brooks, John Waits, and Will Hester. Below, in this 1910 photograph, the two-story building on the left was Charlie Sheffield's store; the second floor housed a millinery shop.

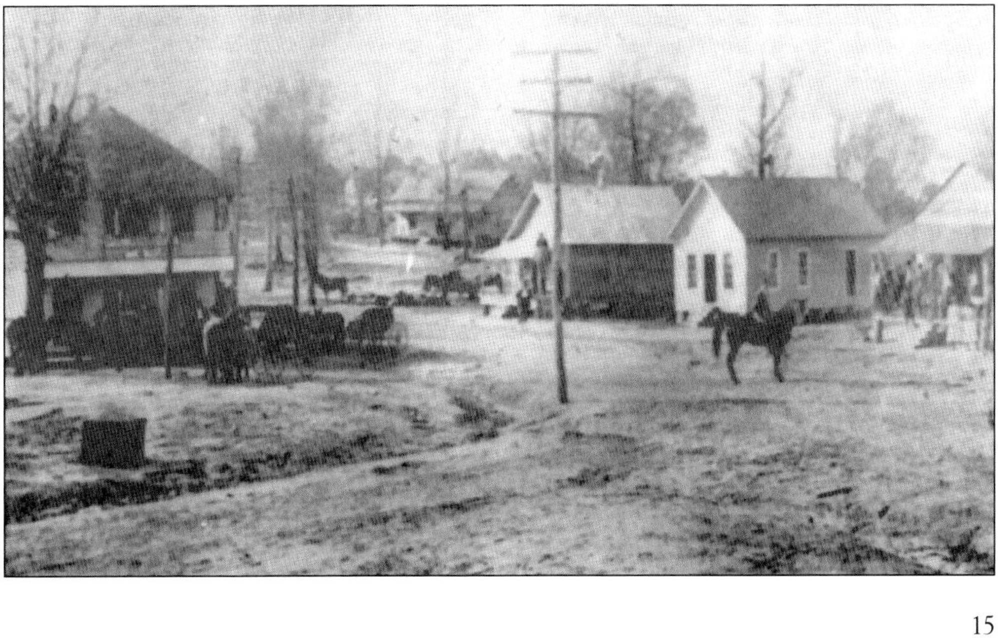

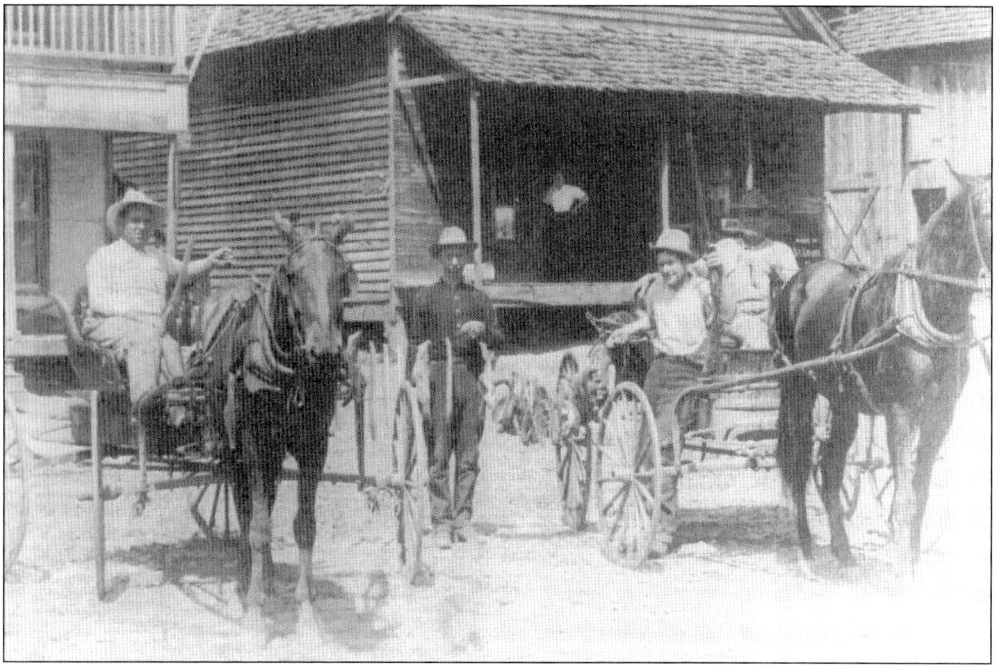

McFerrin's Store was a longtime Mantachie institution. Timothy Woods Pearce was the original proprietor, but Thomas L. McFerrin purchased the store in October 1919 from T.E. Jones. The McFerrin family operated the store nearly 70 years. Pictured in front of the store about 1910 are Al Grissom, mail carrier, with Judd and Audie Turner and their father, Steven H. Turner.

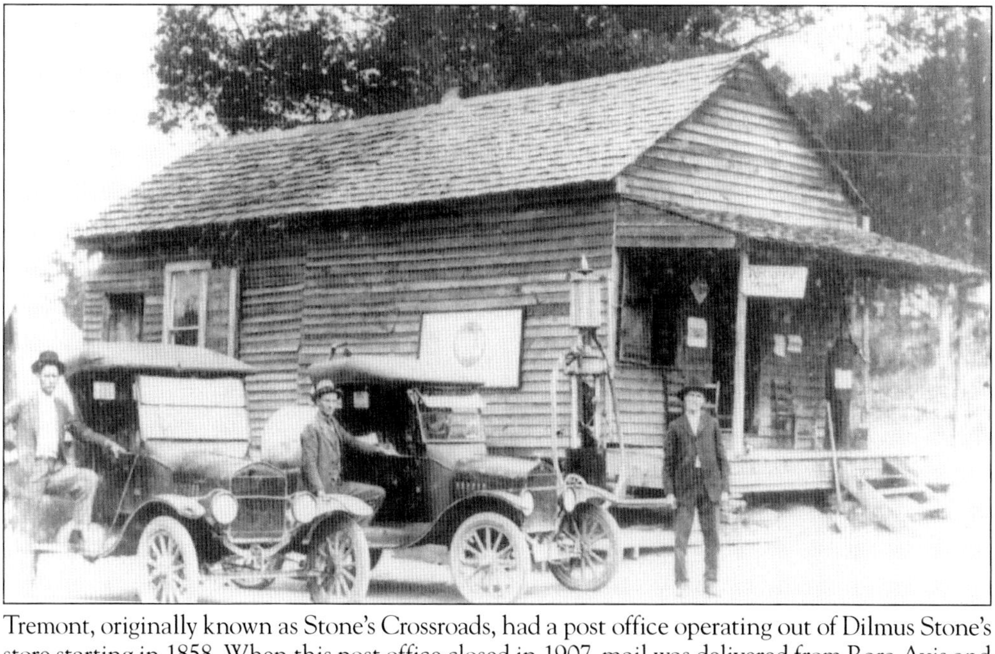

Tremont, originally known as Stone's Crossroads, had a post office operating out of Dilmus Stone's store starting in 1858. When this post office closed in 1907, mail was delivered from Rara Avis and Smithville until 1922 when a post office was reestablished at Tremont. Pictured from left to right around 1925 are Ray Robinson, Owen Spearman, and postmaster M.D.L. Spearman. Oleane Spearman Wright, great-granddaughter of Dilmus Stone, served as postmistress for 35 years, until 1970.

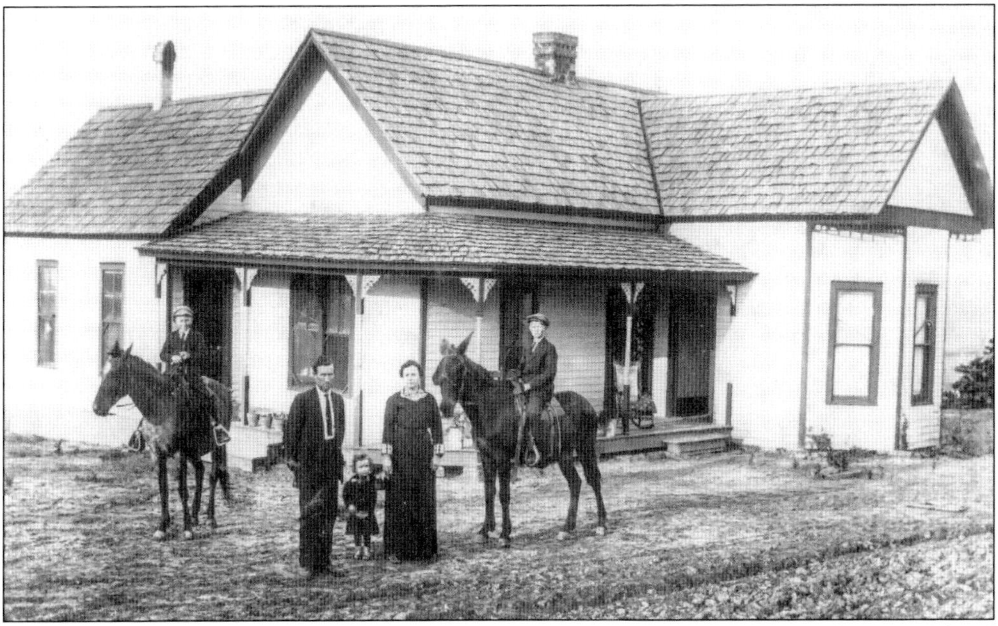

The Carlton McKindrey Robison family lived in a stylish home in Tremont along Bankhead Highway, on land originally settled by Talmon Harbor, Carlton's father-in-law. Myrtle Harbor Robison ran a boardinghouse out of her home. Notable guests included Congressman Ross Collins, for whom Ross Collins Robison was named. From left to right are Roy, Carlton, Ross, Myrtle, and Ray; sons Ruble and Rex were not yet born. (Courtesy of the Ross C. Robison family.)

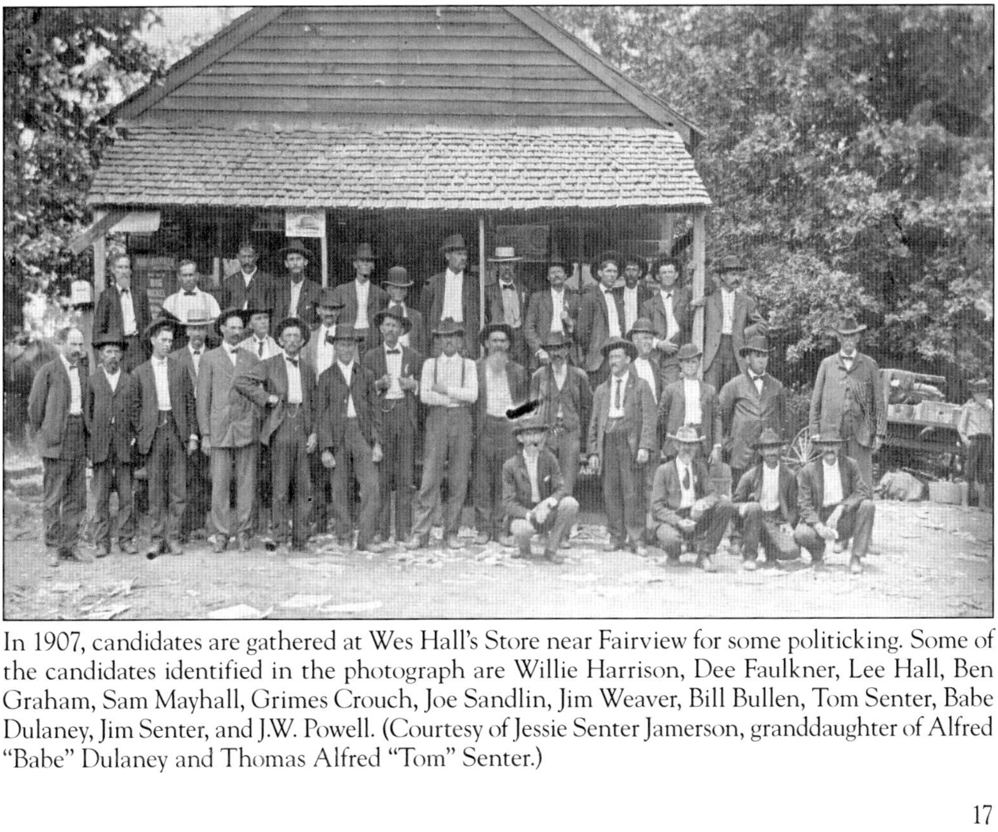

In 1907, candidates are gathered at Wes Hall's Store near Fairview for some politicking. Some of the candidates identified in the photograph are Willie Harrison, Dee Faulkner, Lee Hall, Ben Graham, Sam Mayhall, Grimes Crouch, Joe Sandlin, Jim Weaver, Bill Bullen, Tom Senter, Babe Dulaney, Jim Senter, and J.W. Powell. (Courtesy of Jessie Senter Jamerson, granddaughter of Alfred "Babe" Dulaney and Thomas Alfred "Tom" Senter.)

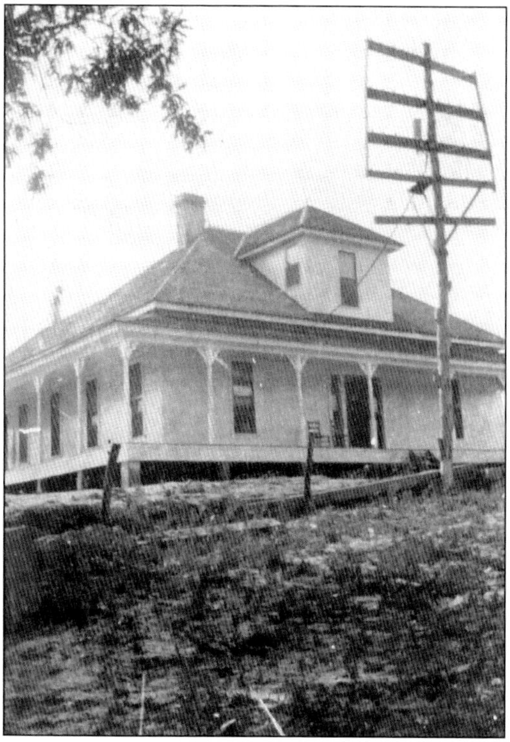

Walter C. Cummings Sr. built this home in the Ratliff community and in 1911 started one of the first telephone exchanges in the western part of the county, using his children as operators. Cummings was the grandson of George Ratliff, founder of the community. In this 1917 photograph, a large telephone pole stands just beyond the Cummings house. (Courtesy of Coleen Thornton, great-granddaughter of W.C. Cummings.)

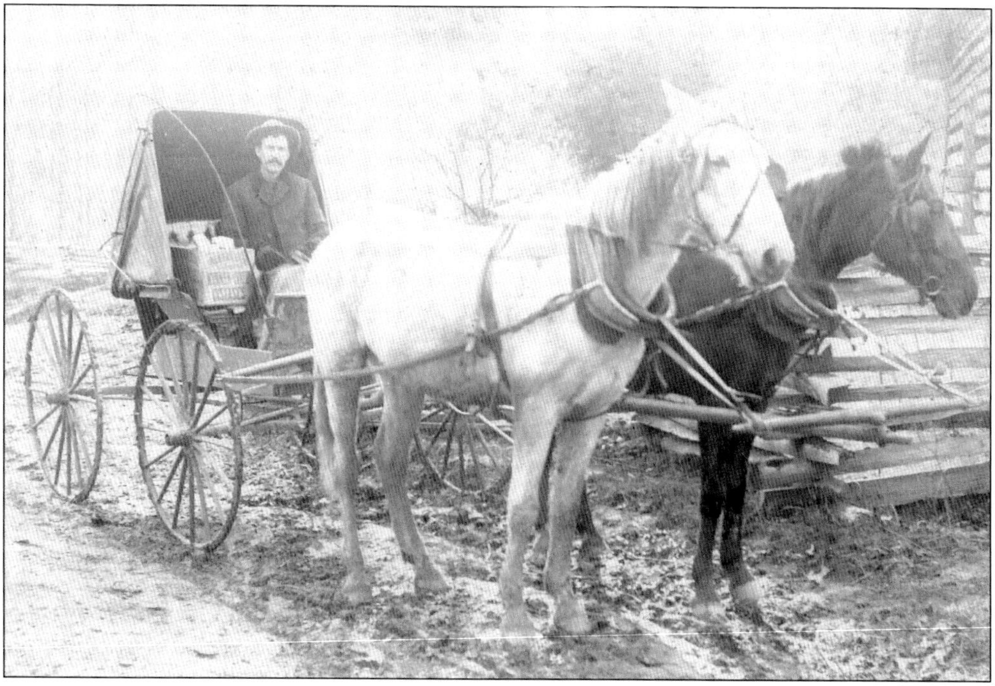

Thomas Daniel Harden, father of *Itawamba County Times* publisher Delmus C. Harden, started the first rural mail delivery route in the county—Route One—out of Fulton in 1901. The mail was first delivered by horseback, but later the elder Harden acquired a buggy and team for the route. (Courtesy of the Delmus Harden family.)

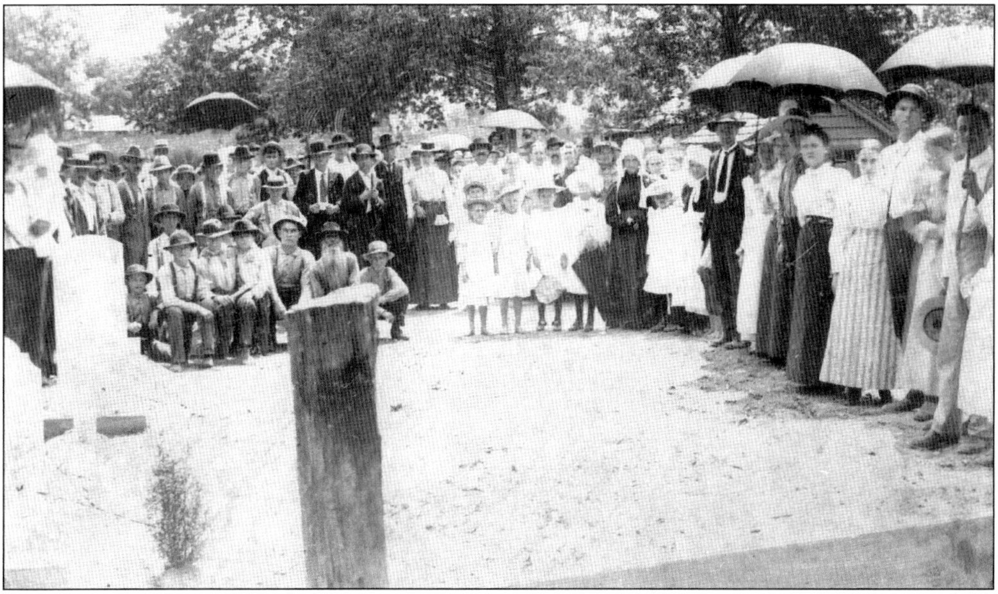

A crowd gathers in a cemetery on a bright, sunny day around the turn of the 20th century. The occasion was probably Decoration Day, an annual event that continues to this day in several cemeteries around the county. Decoration Day usually was held in August, lasted most of the day, and involved preaching, singing, and dinner on the ground. Note the bare ground; in earlier days, cemeteries were swept to remove all grass.

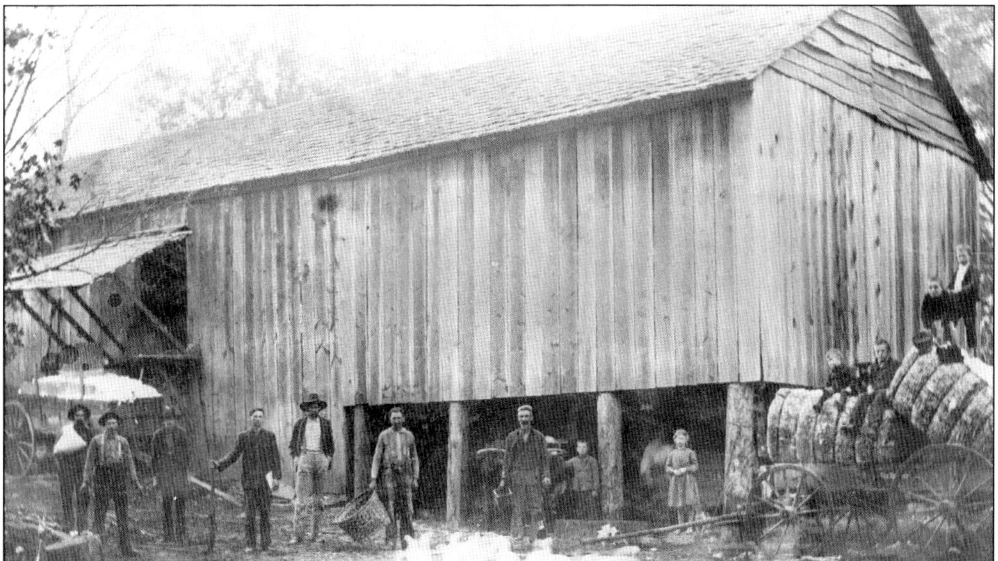

Standing at the mill and gin at Mud Creek about 1904 are, from left to right, Roe Russell, Fisher Johnson, Oscar Adams, Sam Johnson, Harrison Hood, Tom Graham, Henry Johnson, Hess Johnson, Saley Johnson, and Hattie Scott. On the wagon, from left to right, are Noah Adams, Arch Johnson, Eli Johnson, and Todd Russell. William Elisha Bowen, the mill's second owner, was a strict moralist and put a stop to men using the mill to grind their corn for making whiskey.

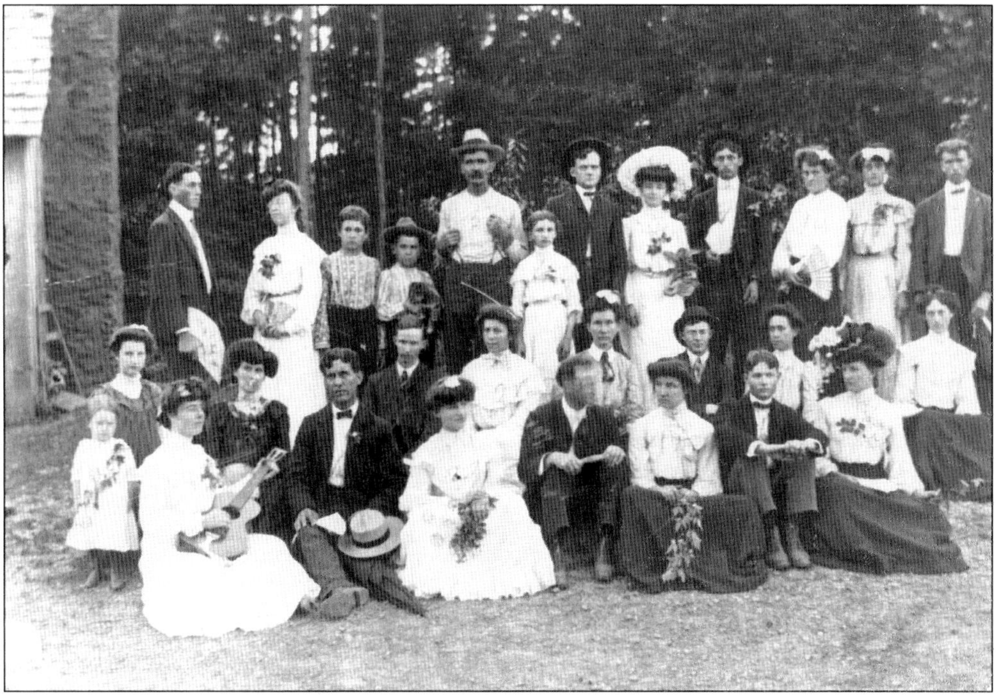

Members of a stylish wedding party pose for a photograph at the old Mt. Pisgah Church east of Fulton. The original Mt. Pisgah Baptist Church was located on the west side of Nita Lake; all that remains is the old cemetery. The tall gentleman in the back is James Louis Bookout; most of the other people remain unidentified.

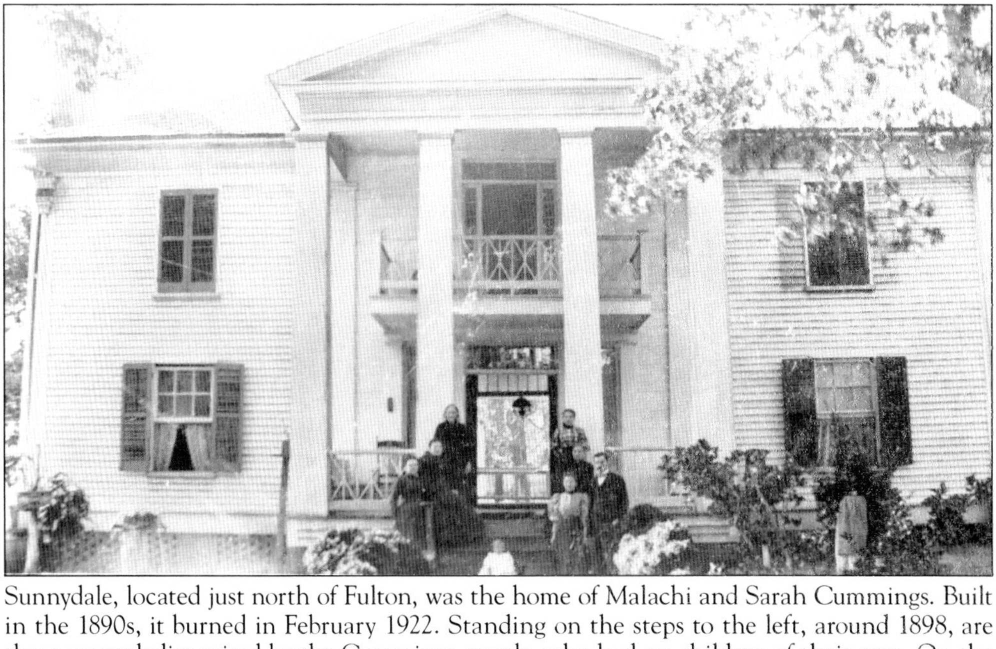

Sunnydale, located just north of Fulton, was the home of Malachi and Sarah Cummings. Built in the 1890s, it burned in February 1922. Standing on the steps to the left, around 1898, are three young ladies raised by the Cummings couple, who had no children of their own. On the right steps, from left to right, are Norma Evans, Sarah Rogers Cummings, Joe Sims, and Nannie Gaither Sims. Yancy Sims is the young boy in the wagon.

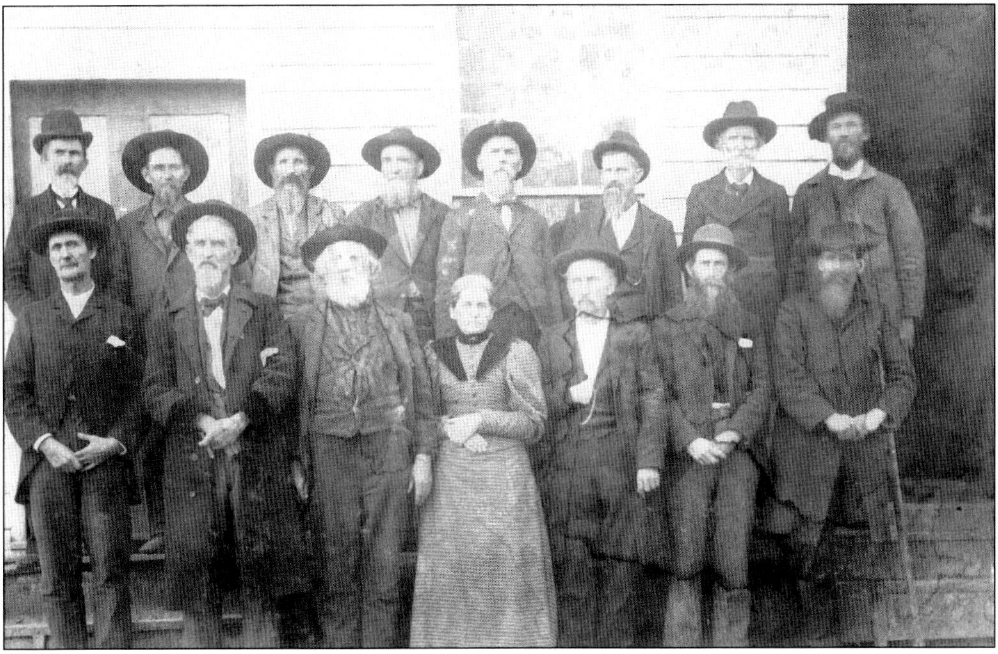

Following the Civil War, Confederate veterans often gathered to relive their shared experiences. Many reunions took place at the county courthouse. They were organized family affairs with music, speeches, fellowship, and food. In 1895, former Confederate soldiers gathered for this photograph; notice one man's missing leg. The woman was perhaps the hostess for this gathering and may have been married to one of the veterans.

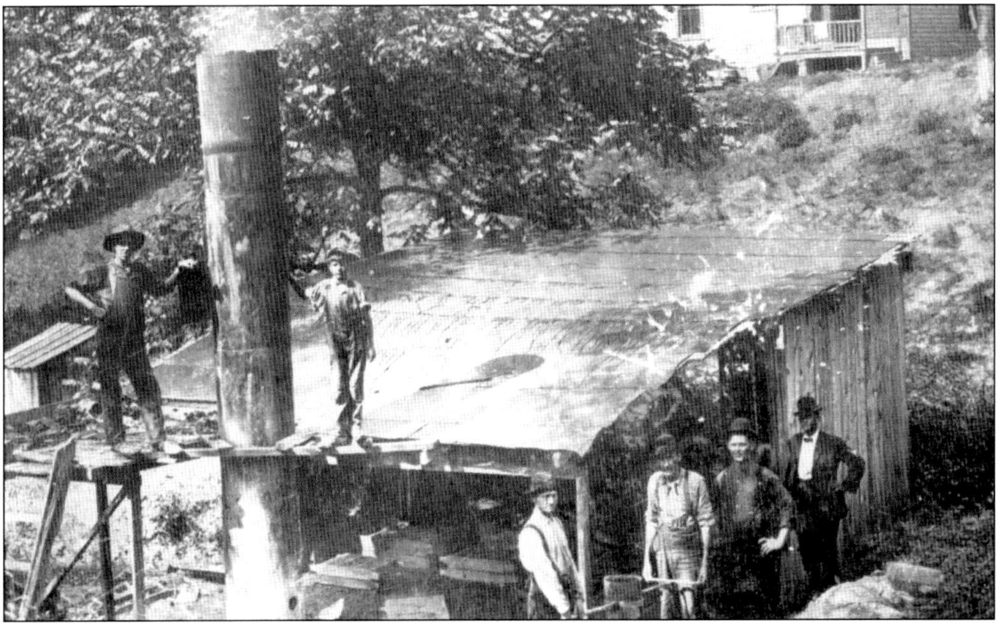

Sorghum cane was grown by most farmers to provide syrup for Itawamba tables. Harvested in the fall, stalks of sorghum were cut, stripped, and taken to a nearby mill such as this one for processing. For those families who did not live near a sorghum mill, there were millers with portable equipment who travelled from farm to farm. Sorghum syrup was important as a sugar substitute.

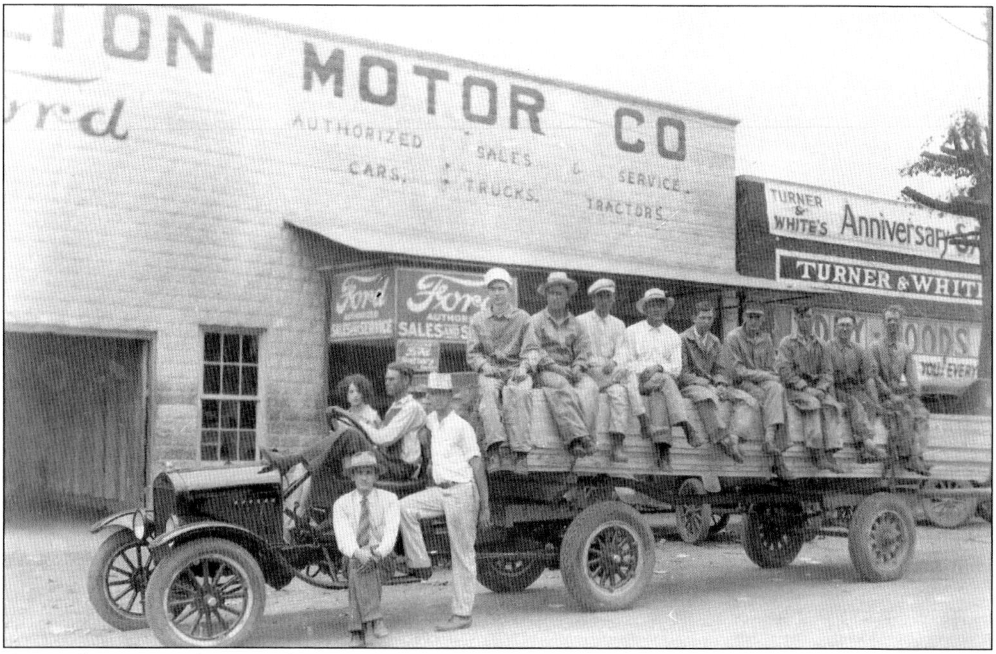

Opened in September 1921 by James M. Brown and L.A. Alsworth, Fulton Motor Company was an authorized dealership for Ford vehicles and Fordson tractors. A common practice for the Ford Motor Company was to caravan new vehicles across the country to its dealerships for public display; it is likely this truck was part of a traveling caravan. Brown and Alsworth are probably the gentlemen posing on the front of the truck while the company's salesmen and mechanics sit on the lumber. As advertised in *The Itawamba County News*, a new truck with pneumatic tires cost $445, while a Ford sedan, with starter, sold for $660. Below, mechanics of Fulton Motor Company take a break from their jobs for a photograph.

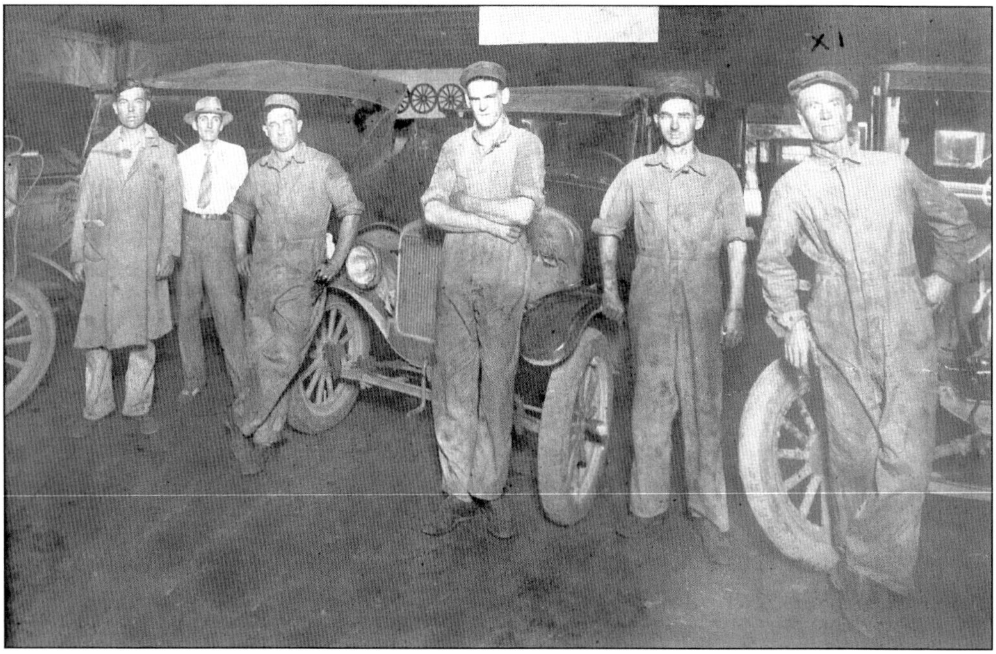

In 1920, Zereda Greene snapped this photograph of the Tupelo-Fulton mail wagon stopped at the Dorsey Post Office. Ordinarily, a mail car handled the route, but the wagon was necessary due to bad roads that particular spring. Will Wiygul, second from left, was the mail carrier that day. W.S. "Bett" Sheffield (left), postmaster at Dorsey, and William A. Martin are standing in front of the wagon.

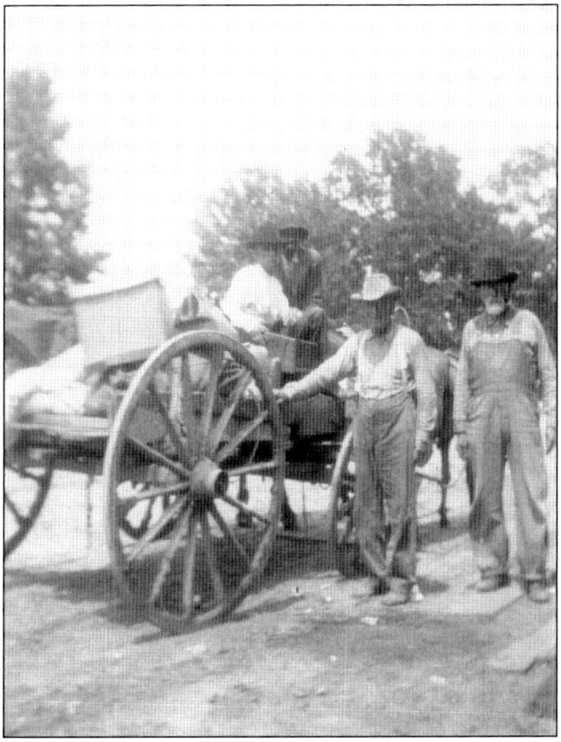

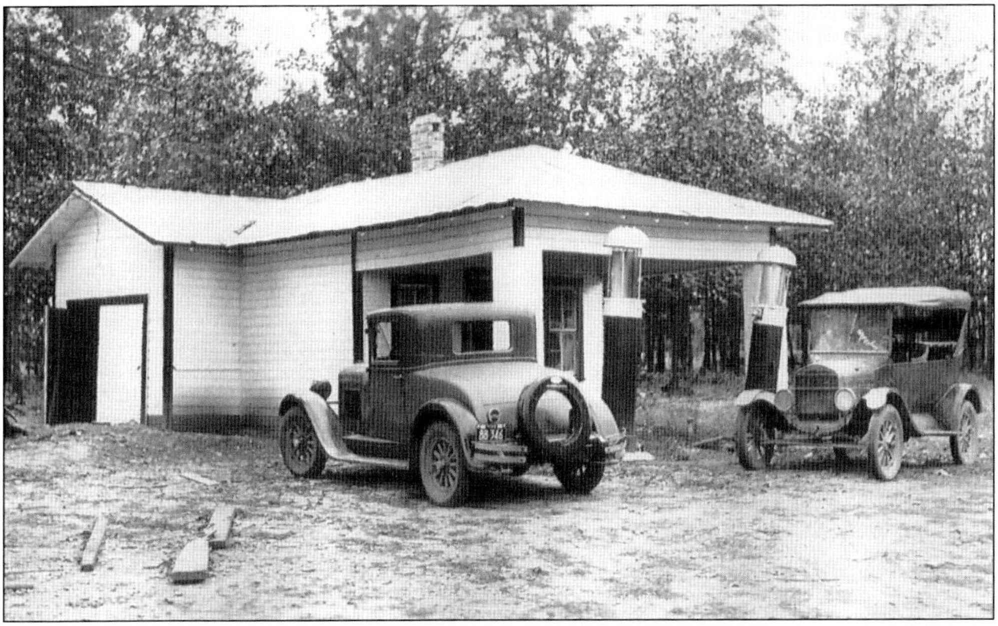

As automobiles and trucks became more popular, so did the need for greater accessibility to gasoline. Noonan Z. Riley operated the Dorsey Service Station to accommodate travelers on Bankhead Highway between Fulton and Tupelo. Before opening the service station, Riley was an automobile dealer at Commerce and Main Streets in Tupelo. He was also well known as a mule and horse trader in Lee and Itawamba Counties. Earlier, he had developed real estate in Tupelo. (Courtesy of Julian Riley.)

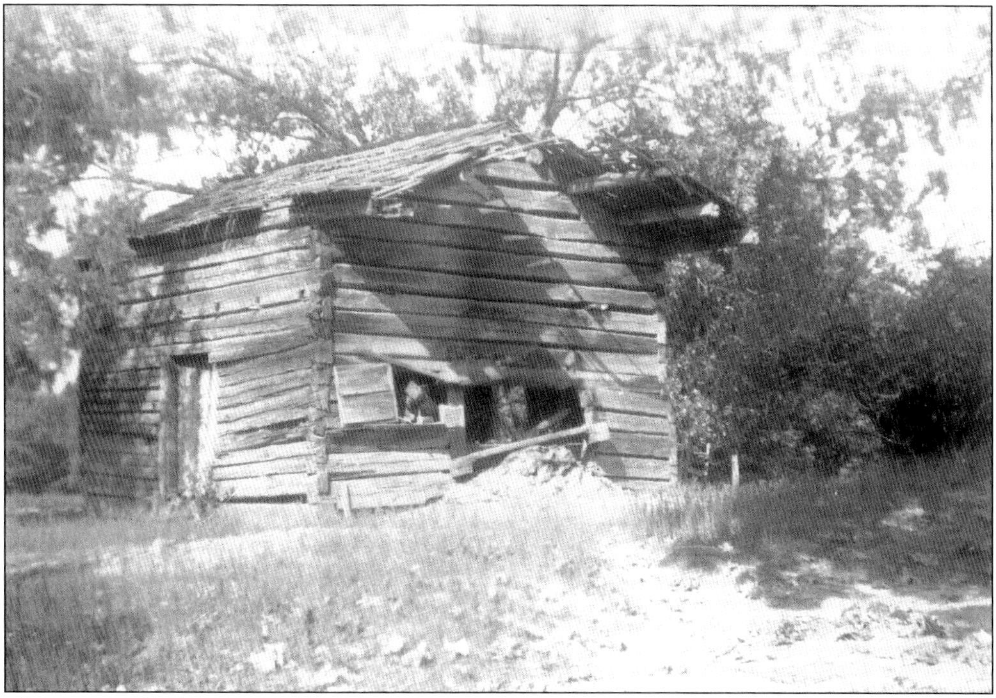

The pioneer log home of Henry and Sarah Johnson was located on the Fulton-Eastport Stage Route. Built in the 1830s by Mac Shaw, the home and associated land was purchased by Johnson in 1851 from Thomas Winkler, a subsequent owner. The house was constructed using ridgepoles instead of rafters, a unique feature. Johnson, a North Carolina native, married Sarah Ingle, of Indian ancestry, in 1842 in Itawamba County.

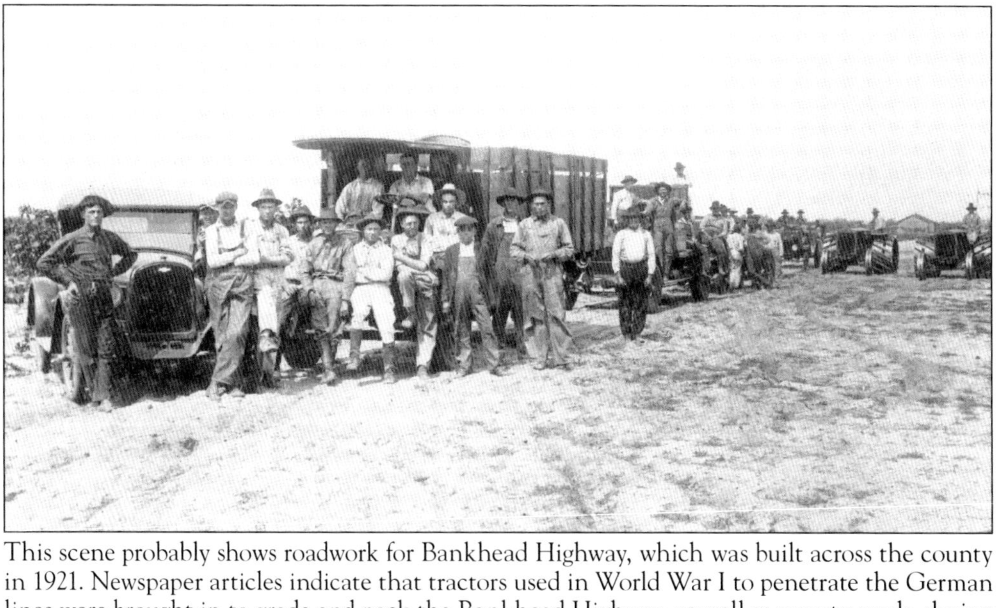

This scene probably shows roadwork for Bankhead Highway, which was built across the county in 1921. Newspaper articles indicate that tractors used in World War I to penetrate the German lines were brought in to grade and pack the Bankhead Highway, as well as county roads, during that year. The road from Fulton Cemetery west to Pontotoc was surfaced with concrete, while a hard-surfaced gravel road connected Fulton to the Alabama state line.

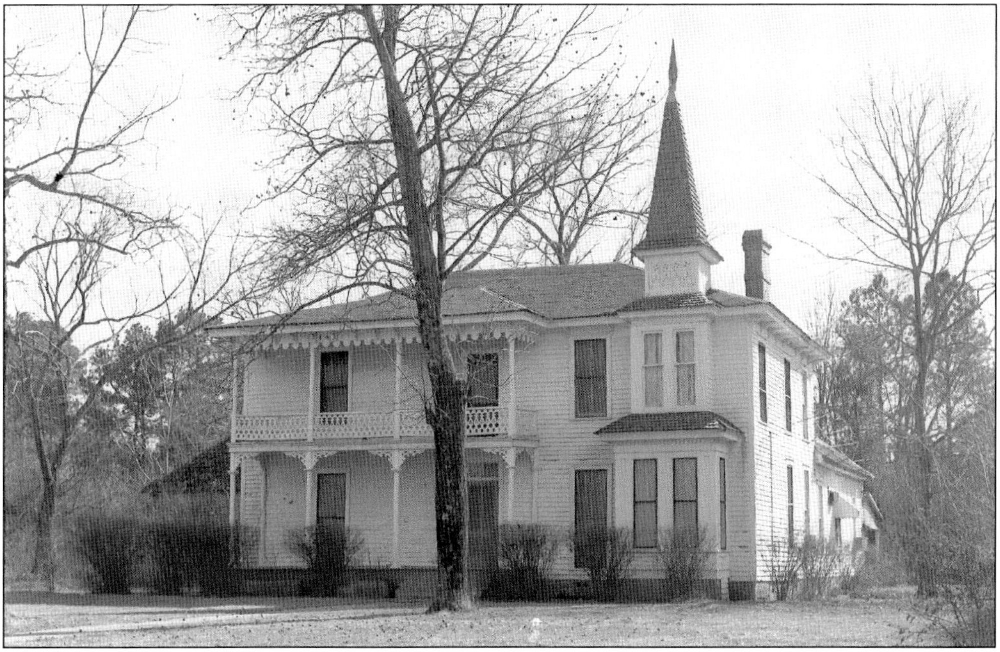

Known to most folks as the former Orr home, this house in Fulton was built around 1840 by a Kinsolving, of Swedish descent. Pleasant Cates bought the house but later sold it to his son-in-law, J.P. Baldridge, who added rooms and a second story to the home. In 1903, his widow, Elizabeth Cates Baldridge, sold the house and all its furniture to W.G. Orr, local merchant and father of Dr. William Lawrence Orr.

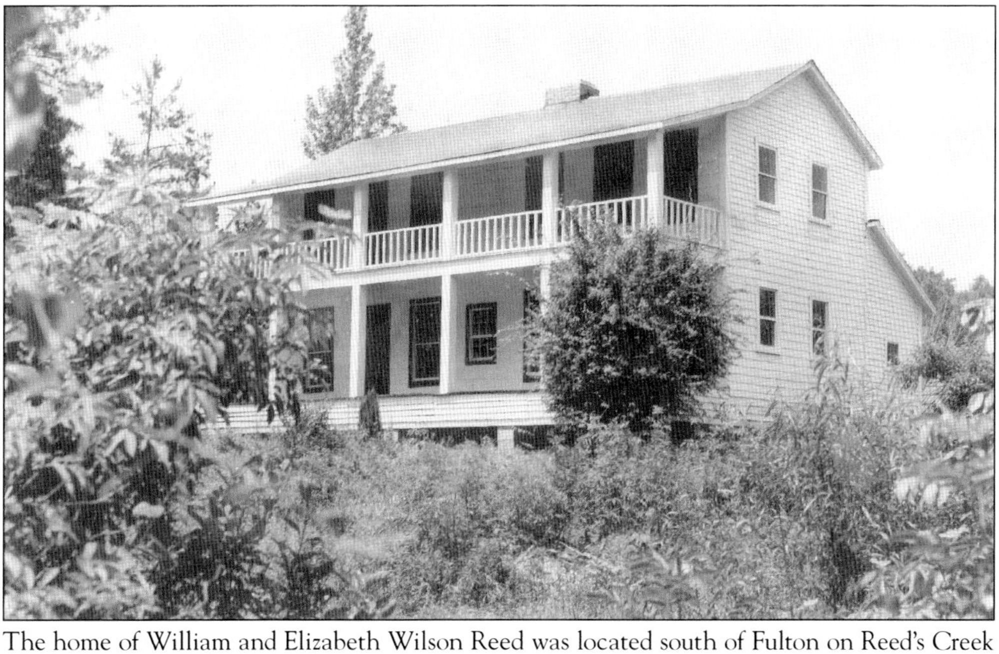

The home of William and Elizabeth Wilson Reed was located south of Fulton on Reed's Creek below Bean's Ferry on the Tombigbee River. William, a South Carolina native, owned a mill as well as more than 1,000 acres of land. William and Elizabeth raised a large family of 15 children, and their many descendants populate the county today.

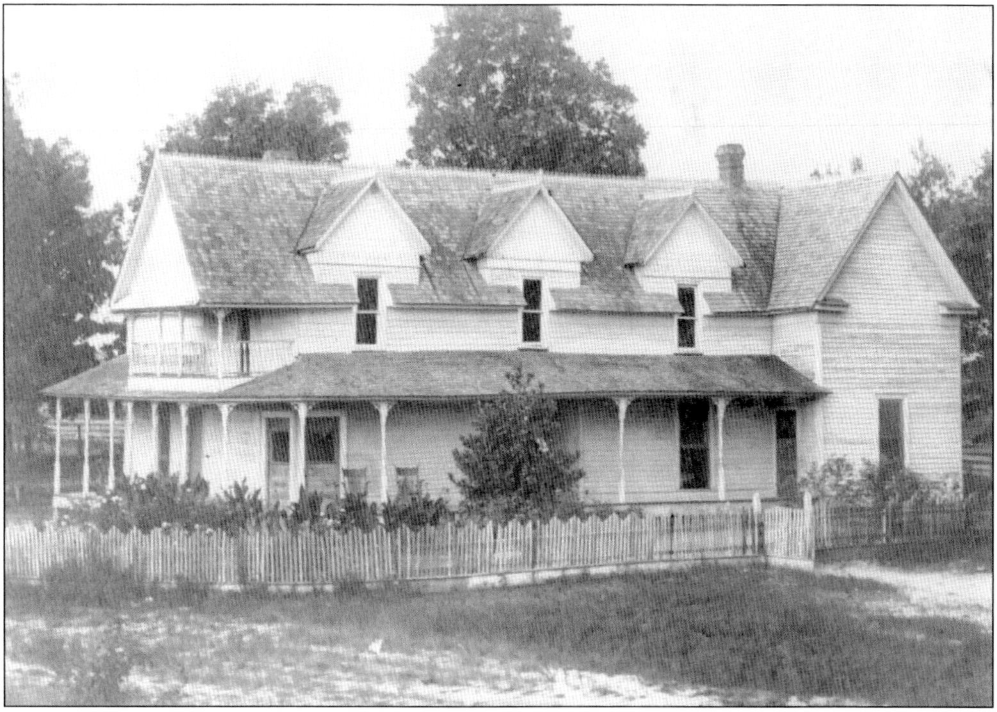

Robert Sidney and Leona Gregory Sheffield built this home around 1900. Located in the Dorsey community, it featured 12-foot ceilings, a dogtrot, porches on all sides, a balcony, and dormer windows upstairs. The home was torn down in the 1950s. Leona and Robert had nine children; he was a prominent educator and represented Lee and Itawamba Counties in the state legislature. (Courtesy of Joe Nell Wood.)

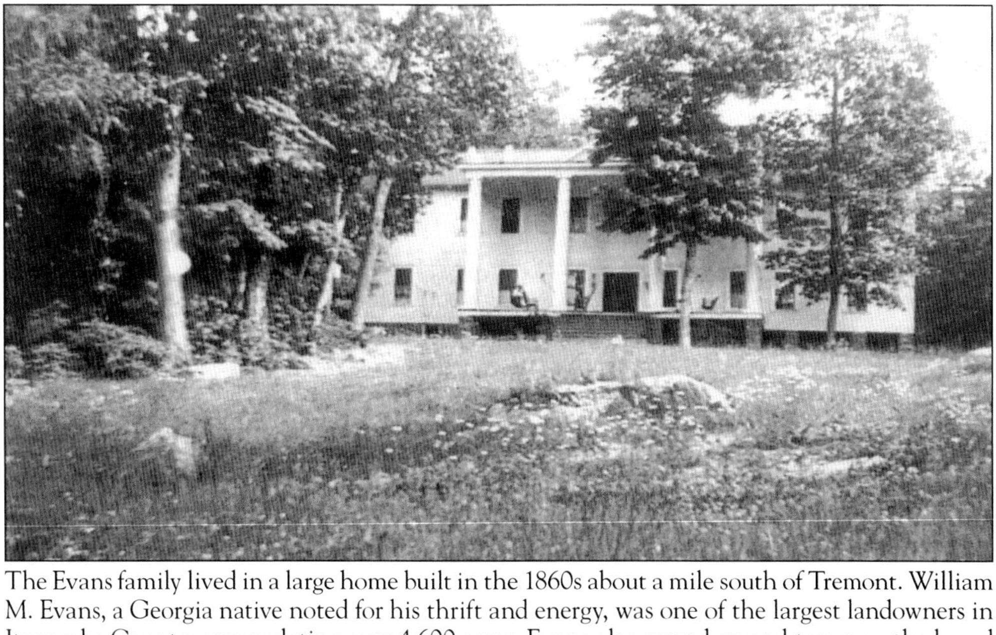

The Evans family lived in a large home built in the 1860s about a mile south of Tremont. William M. Evans, a Georgia native noted for his thrift and energy, was one of the largest landowners in Itawamba County, accumulating over 4,600 acres. Evans also served several terms on the board of supervisors. His wife, Sarah, was the sister of Timothy Woods Pearce of Mantachie.

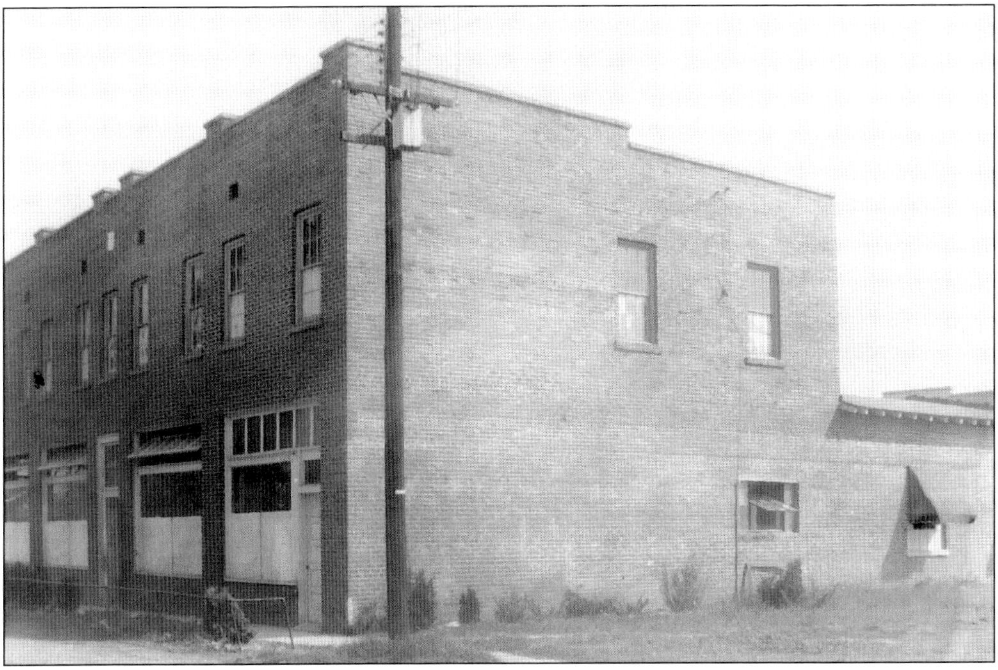

The Fulton Garment Factory, shown above in 1937, was located where the BancorpSouth building is now. It was the only manufacturing plant in the county for many years. Originally, this location on the corner of Wiygul and Cummings Streets was the site of a Ford dealership. Blue Bell Inc. purchased the plant around 1940, eventually employing over 220 workers and outgrowing its downtown location. A new manufacturing plant was built in 1952 on a lot sold by A.T. Cleveland behind his house. In 1966, the old location was demolished to make way for the new Fulton Bank building. Below, employees partake in a Christmas luncheon at the new building. (Above, courtesy of Bob Franks; below, courtesy of the Delmus Harden family.)

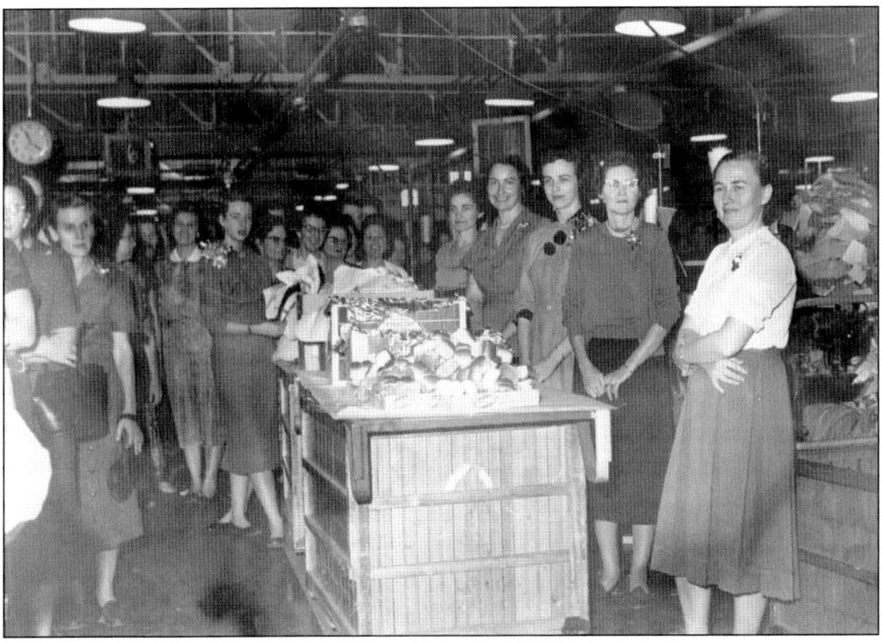

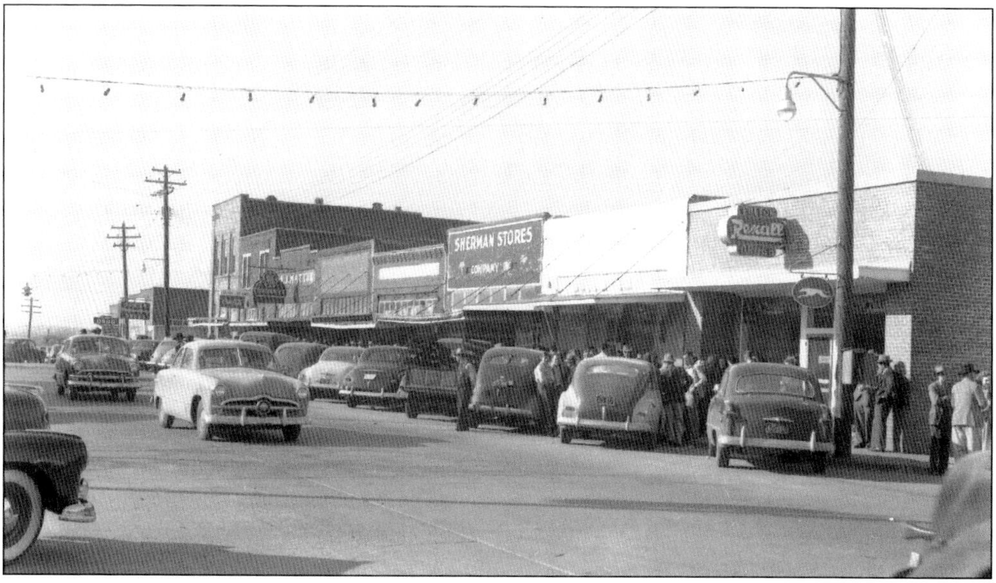

In this photograph from December 1950, Fulton's Main Street appears to be a beehive of activity. The people in front of Fulton Drug Store on the corner are probably waiting for the Greyhound bus that departed from that location. Note the festive light bulbs that are strung across the street from pole to pole at both intersections. (Courtesy of the Delmus Harden family.)

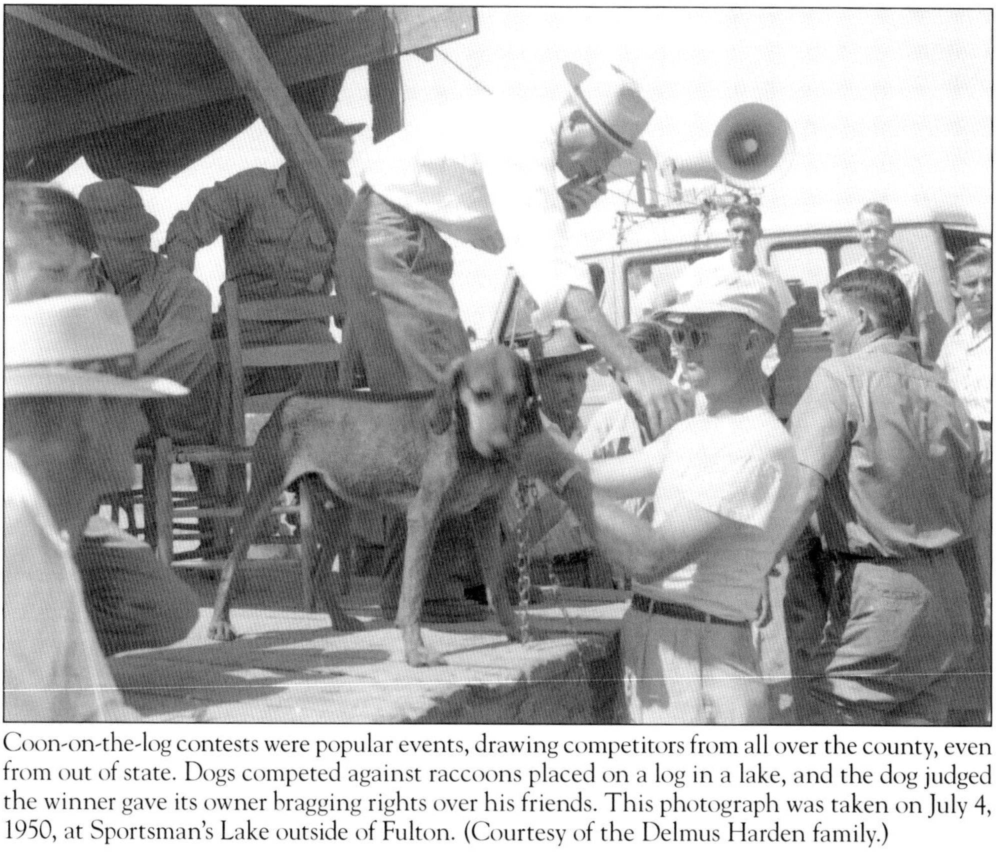

Coon-on-the-log contests were popular events, drawing competitors from all over the county, even from out of state. Dogs competed against raccoons placed on a log in a lake, and the dog judged the winner gave its owner bragging rights over his friends. This photograph was taken on July 4, 1950, at Sportsman's Lake outside of Fulton. (Courtesy of the Delmus Harden family.)

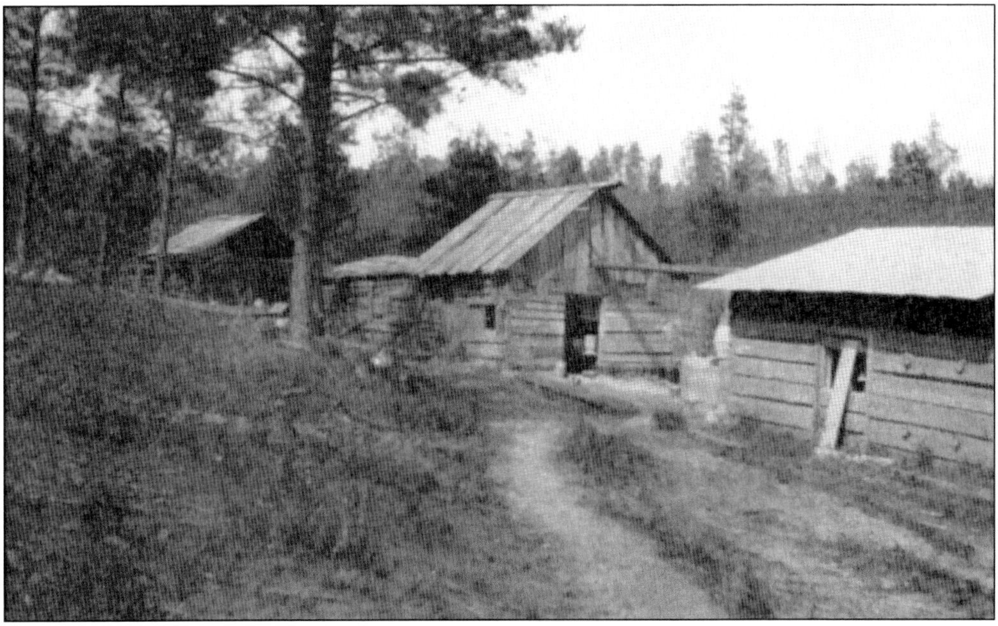

The clay found in the eastern hills and creek banks of Itawamba County was used by local pottery shops to produce thousands of churns and other stoneware for Southern families. John Young's shop (above) used clay dug locally from nearby exposed outcrops to produce its stoneware. An 1888 Fulton newspaper indicates that the hills around James Creek were "dotted everywhere with jug factories" and that the money brought in from the sale of stoneware was "truly astonishing." D.E. Summerford's shop (below) was located in the Cherry Tree community. The stamp on Summerford's stoneware indicates an address of Bexar, Alabama, but his shop was located in Mississippi; Bexar was merely the mail route. Fierce competition and increased use of plastics contributed to the demise of most pottery shops following World War II. (Both, courtesy of *Bulletin: Mississippi State Geological Survey, No. 64.*)

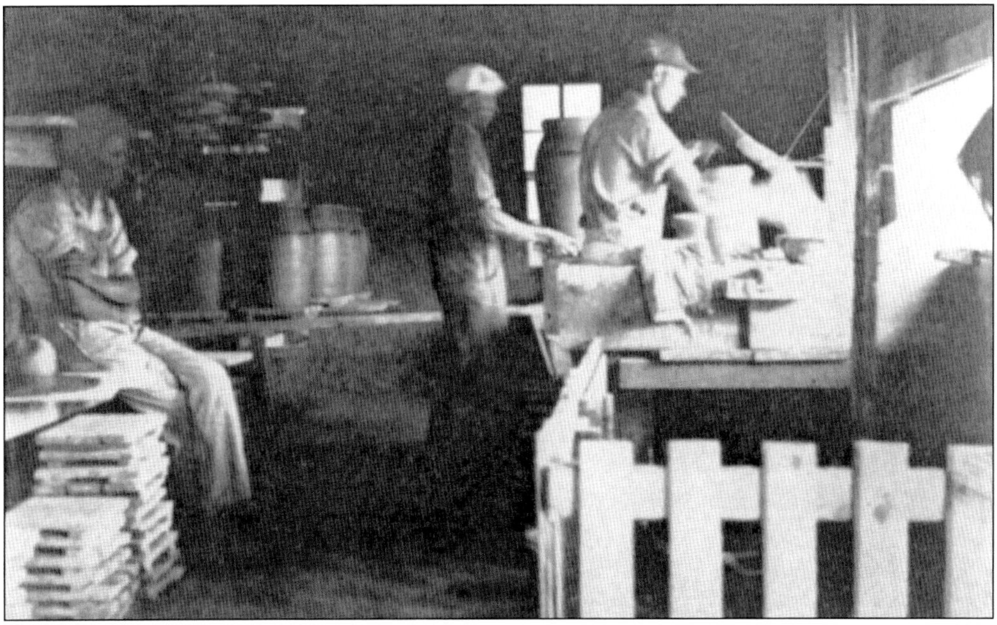

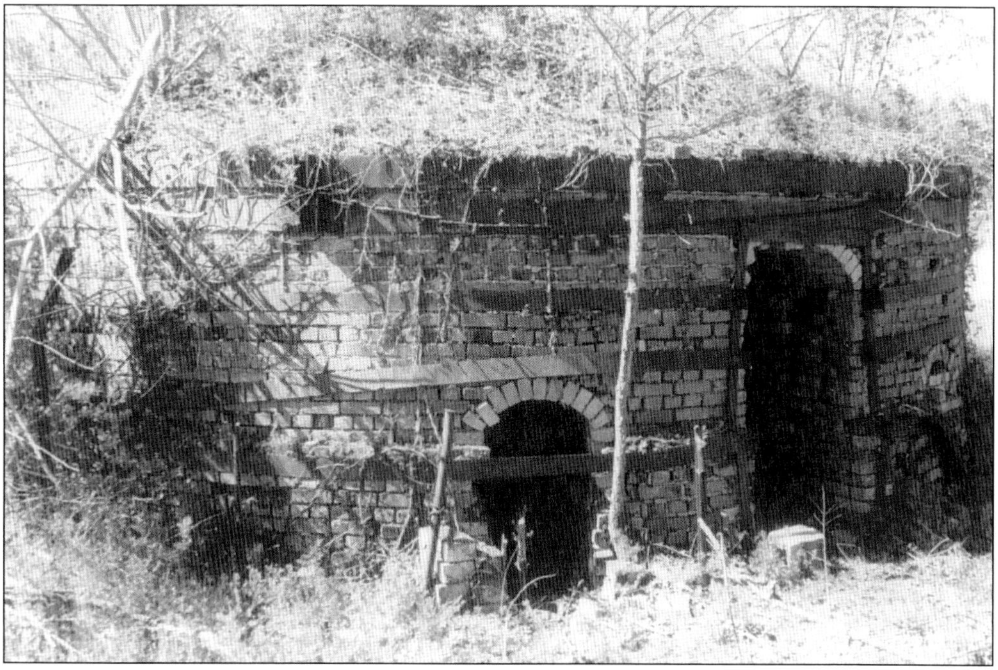

W.D. Suggs established his first pottery shop on State Line Road in the 1890s but later moved to the Turon community north of Smithville, just barely outside Itawamba County. The second Suggs kiln still stands off Jug Shop Road. A circular brick structure enforced with iron bands, this beehive kiln was quite large and enabled the operation to produce large quantities of stoneware. Suggs's shop was one of the larger producers of pottery ware. (Courtesy of James Cormany.)

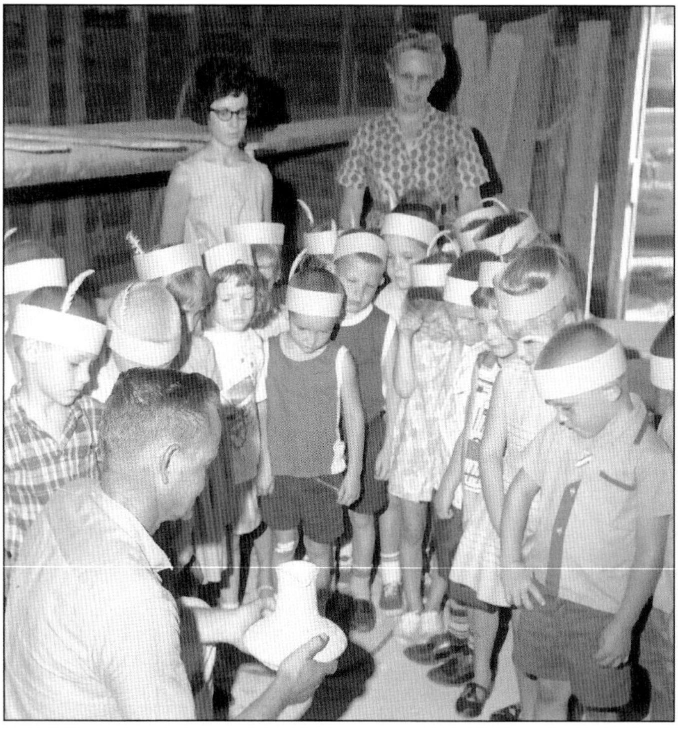

Jesse Harris, son of Solomon Harris, opened his own pottery shop across the Tombigbee River west of Fulton. This shop was located just off Highway 78 at the top of "River Hill" and afforded customers greater access to stoneware made locally. The pottery shop was a popular field trip; in 1965, students from Dorsey Headstart watch as Jesse Harris demonstrates his pottery. (Courtesy of the Delmus Harden family.)

William Payne Loyd and his son William Dickinson Loyd invented and patented this unusual, tablet-style grave marker in 1879. William Payne, the eldest of five brothers, came from a family of potters who were making pottery as early as 1850 in Marion County, Alabama. Loyd's markers were an economical choice for families, who often could not afford larger tombstones for their loved ones. Most of the tombstones found locally were produced by the Suggs shop under license from the Loyd family, with the base stamped "Patented June 10, 1879." Unfortunately, the clay markers have not held up well in the era of lawnmowers and weed eaters, although the cobalt blue ink used for the lettering looks as new as ever. Markers produced directly by the Loyd family can be distinguished from the mass-produced, licensed versions. An original Loyd grave marker almost always contains a leaf or flower drawn either at the top of the tablet or sometimes on the base. This marker in an Itawamba County cemetery was made for William Payne Loyd's son, probably by William himself, and contains a poem written by the deceased's bereaved wife.

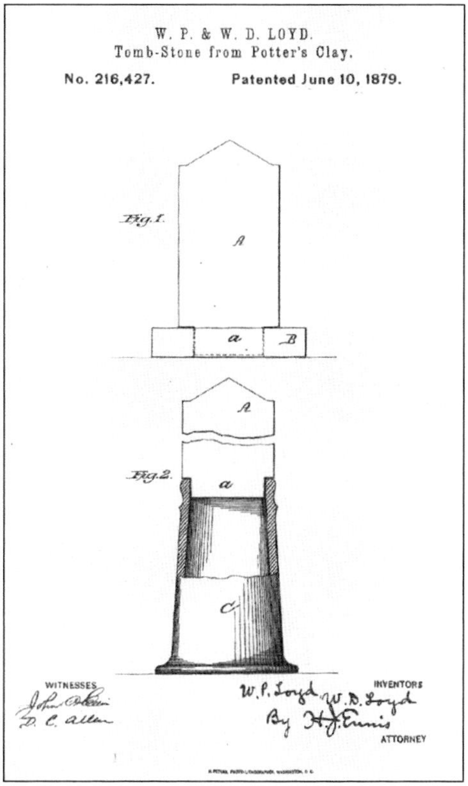

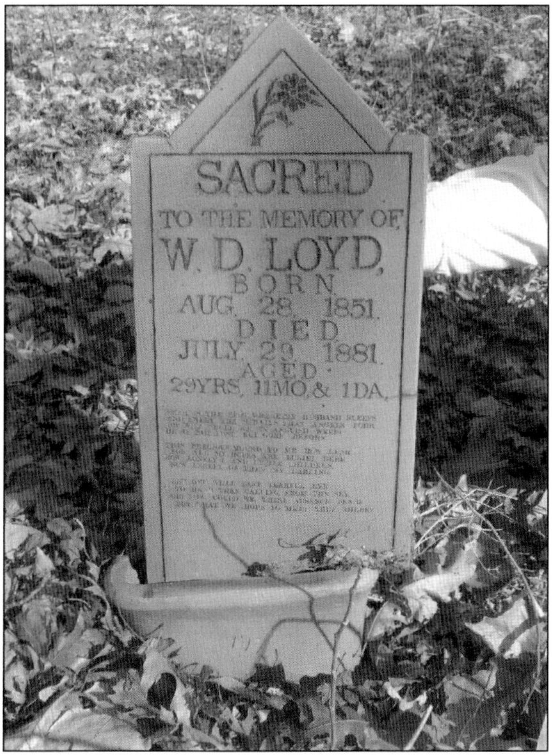

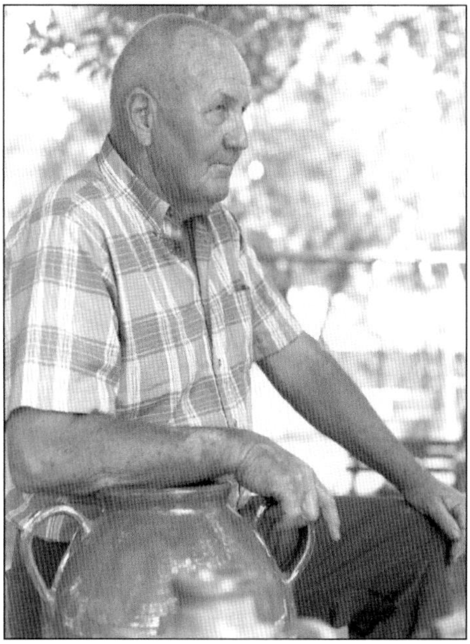

Titus and Euple Riley continued the pottery-making tradition in Itawamba County when they opened Peppertown Pottery in 1983 near the banks of the Tombigbee River. Self-taught, the Rileys dug their clay locally and made their own glazes to create both art and utilitarian stoneware using their hand-built equipment. The Rileys were recognized both locally and nationally for their unique talent. Here, Titus Riley is pictured with some of his clay pottery. (Courtesy of Charles Riley.)

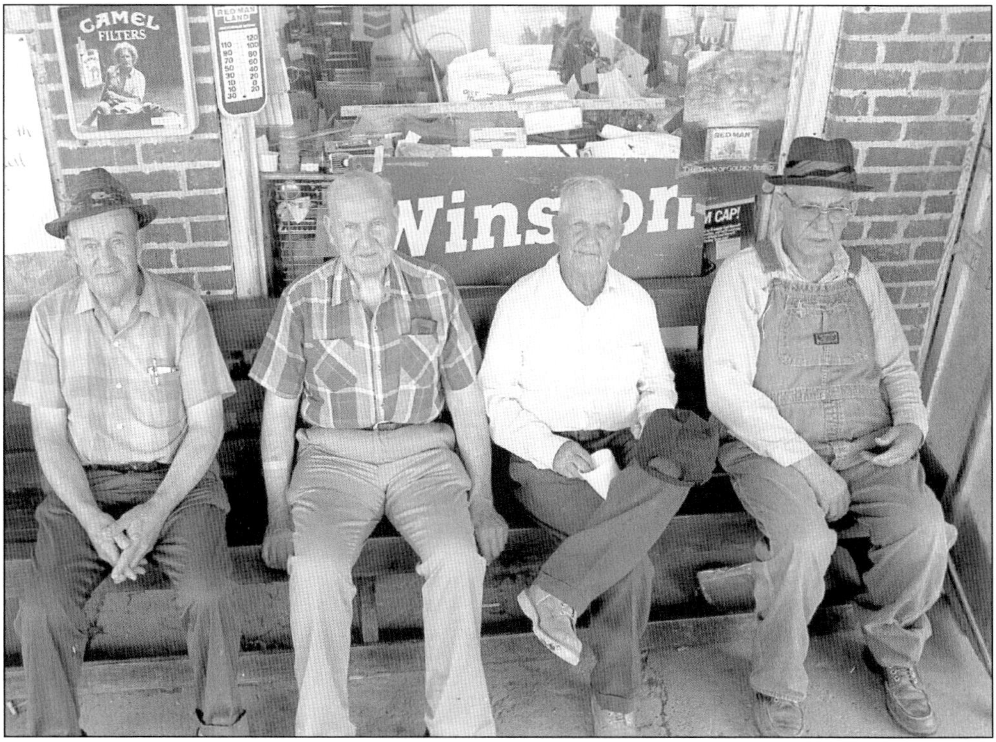

McFerrin Brothers Store was *the* local gathering place in Mantachie until destroyed by fire in 1988. Visiting outside the store in 1987 are, from left to right, Keyron McFerrin and Gordon McFerrin, two of three brothers who inherited the store from their father; Earl Kitchens; and Willie Bill McFerrin. In addition to selling just about anything a shopper could want, the store served as a community bank and carried many people through the Great Depression. (Courtesy of the Delmus Harden family.)

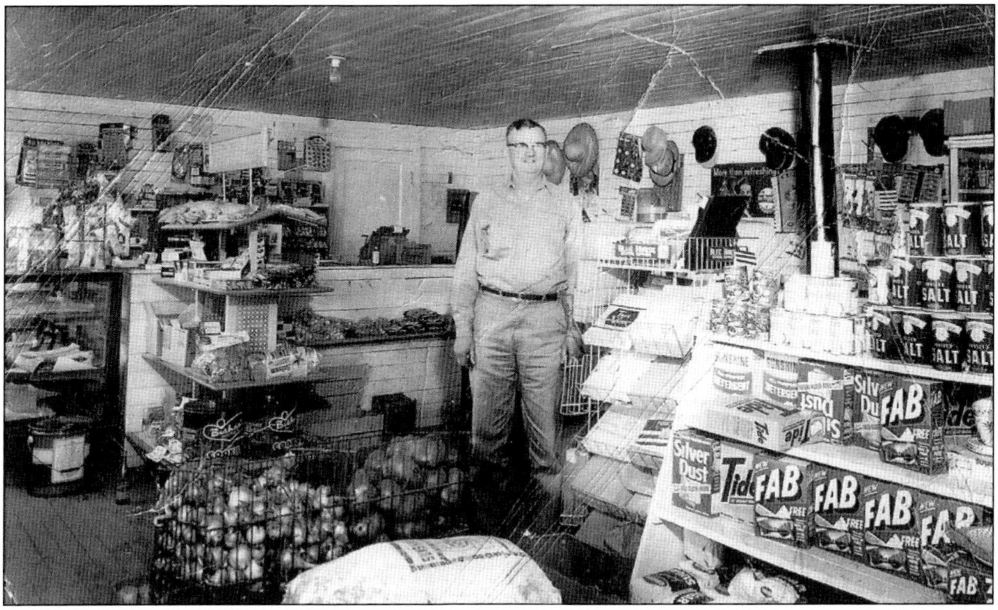

Bennie Cecil Senter was for many years the proprietor of a grocery store located "up the North Road" in Itawamba County. Senter's Grocery was a popular stop for folks traveling between Fulton and the northern reaches of the county. Cecil was the son of James Robert "Jim" Senter and Alberta Chilcoat Senter and the great-great-grandson of Alfred Aven Senter, one of the earliest settlers of the county.

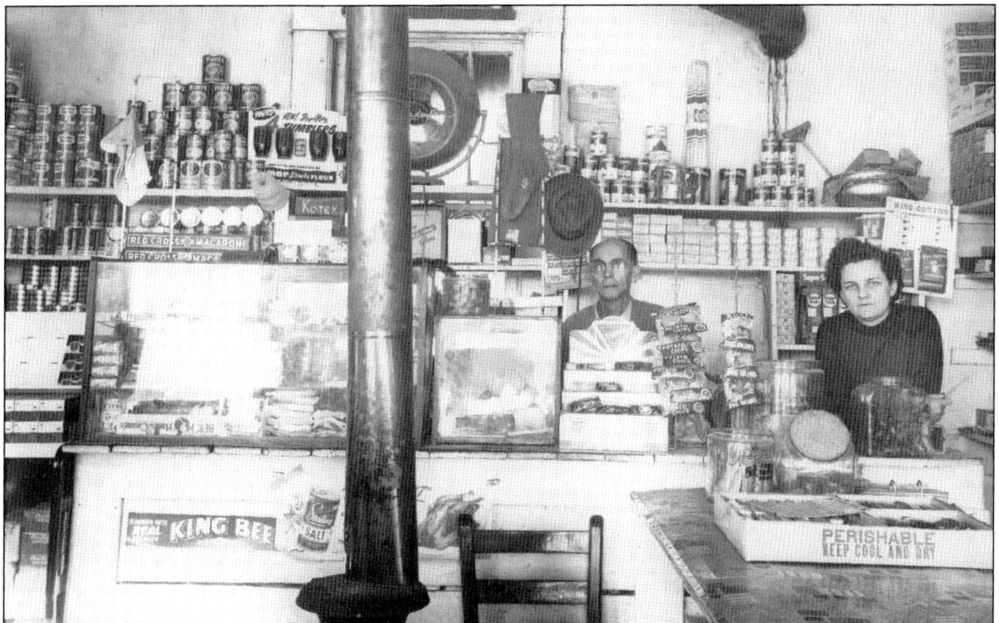

Malburn Laurice Robinson and his wife, the former Sue Wax, operated Tremont Grocery Store. His brother, Mautimer Dewitt Robinson, owned a competing store directly across the street. These brothers were sons of George Emerson Robinson and Virginia Downum. Laurice was a merchant in Tremont for over 50 years. (Courtesy of Virble Booth.)

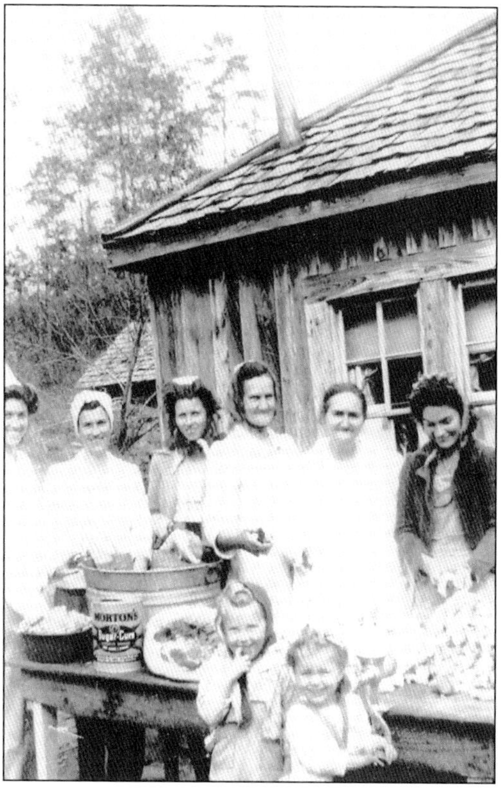

Hog-killing time in Itawamba County created a lot of work for the entire family. Pictured here are members of the Gainey Sloan family of the Peaceful Valley community on a hog-killing day. Keen eyes will spot a ham on the table as well as a sausage grinder. The mound of fat on the right will be made into lard, which at one time was a staple for every southern kitchen. Below, Fisher Johnson of the Ryan's Well community is standing beside one of his prized hogs that had been fattened up with a special diet for about a month before being slaughtered. A fatter hog meant more meat and also more lard. Hog-killings, usually a daylong event, took place in the late fall, when a stretch of cold weather was usually guaranteed. (Left, courtesy of Gail Newton Carter.)

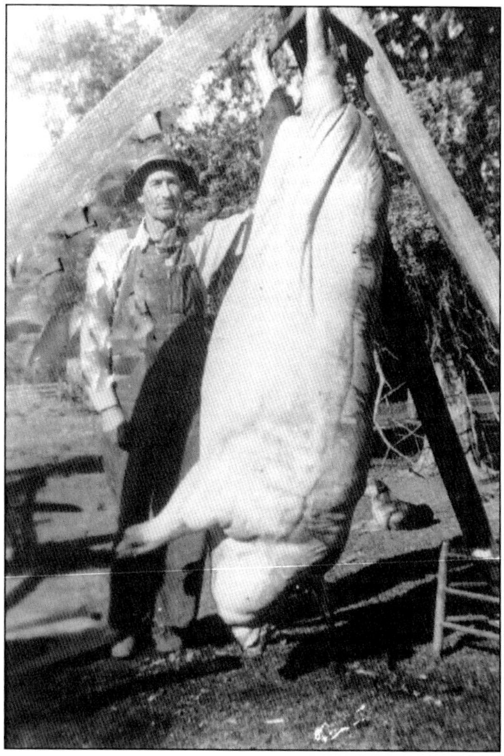

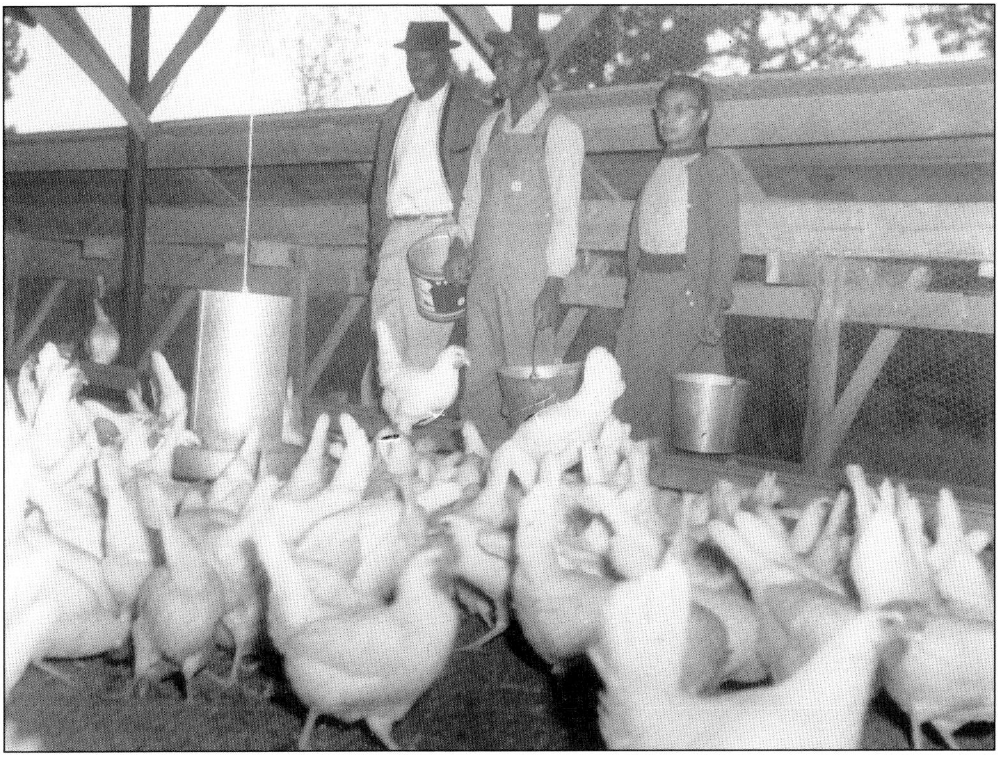

Troy and Annie Sargent of the Carolina community were 1959 winners in the Plant-to-Prosper program sponsored by the *Itawamba County Times*. In addition to having 47 acres in cultivation, their farm included two "laying hen" houses and a Grade A dairy. Pictured with the Sargents is C.C. Cummings, left, vocational agriculturalist for the county. (Courtesy of the Delmus Harden family.)

The first ginned cotton bale of the season was a photo opportunity every year. Oscar Burl Barnes, second from left, owner of the Ozark Gin, hands a check to an unidentified customer. Seated to the left is onlooker G.C. Spigner. In addition to running a cotton gin in the Ozark community, Barnes also was a farmer. (Courtesy of the Delmus Harden family.)

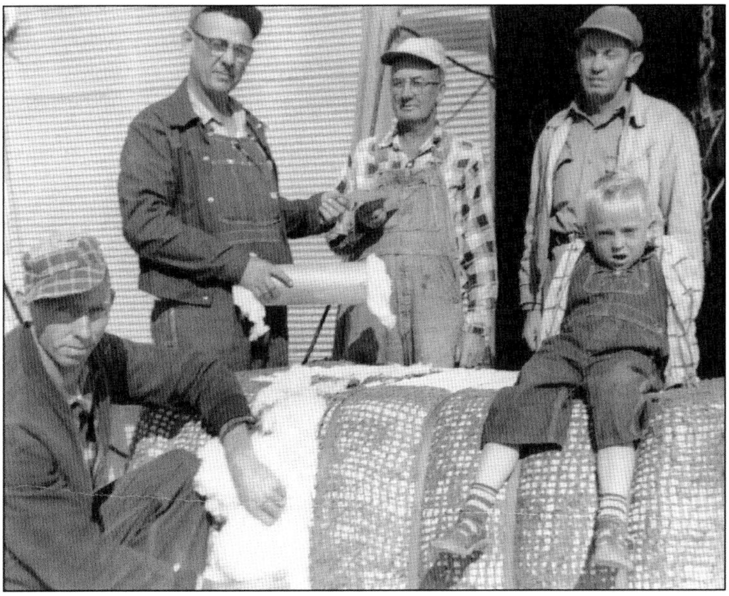

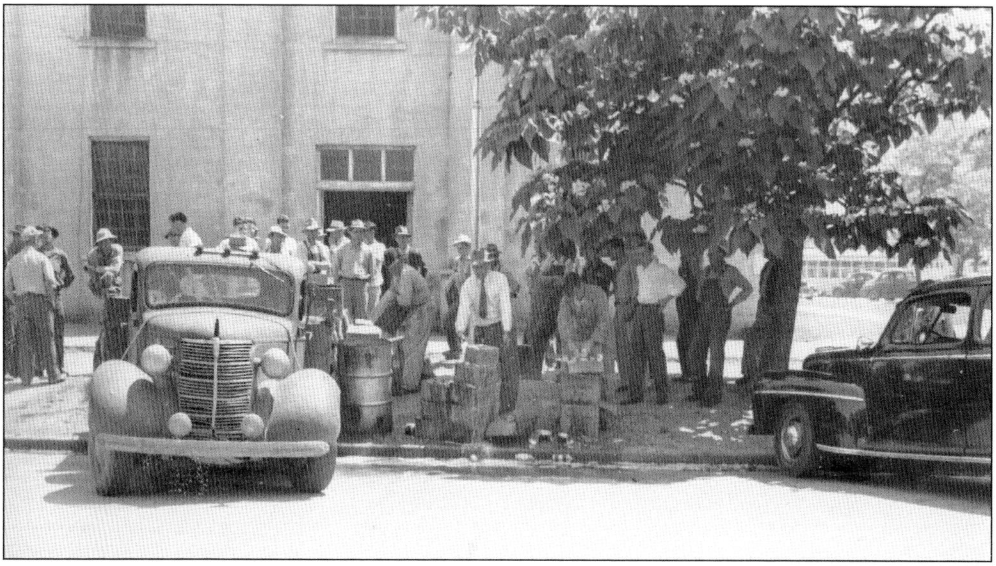

Moonshine stills existed in Itawamba County for most of the 20th century, although eventually moonshining operations died out to be replaced by bootlegged beer and whiskey, both illegal products in a dry county like Itawamba. The courthouse lawn was a good place for the public to witness the sheriff doing his job, especially when election time was nearing. (Courtesy of the Delmus Harden family.)

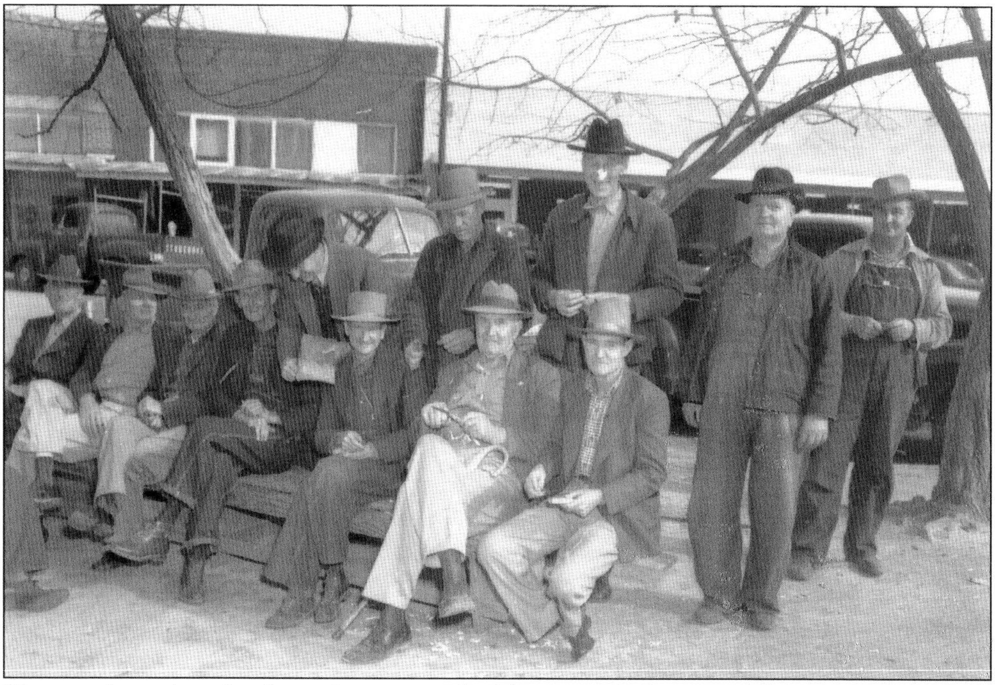

The courthouse lawn has been the most used public site in the history of the county. Everything from knife swapping, as shown here in the 1950s, to horse trading (see page 10), preaching and singing, parades of dairy cattle, farm equipment exhibitions, art shows, candidate speakings (see page 11), festivals, and a whole lot more has taken place on the lawn of the courthouse. (Courtesy of the Delmus Harden family.)

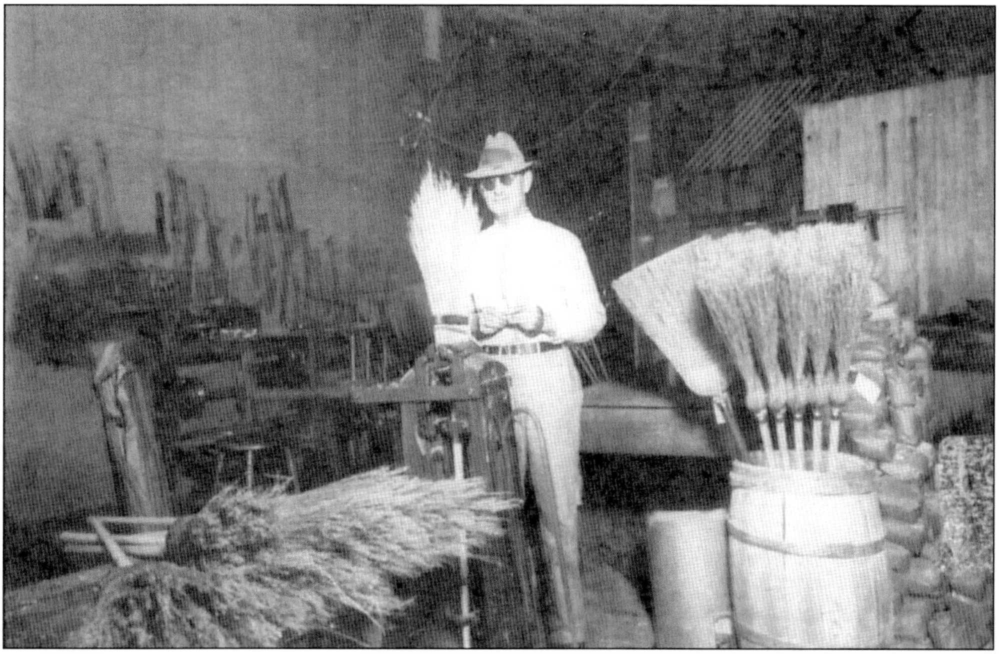

After becoming totally blind in 1932, Henry Hankins learned the craft of broom making. Hankins owned a broom manufacturing business in downtown Fulton, making and selling straw brooms to Itawambians for several years. He was known for his determination and cheerful disposition as well as for his ability to recognize and greet most fellow citizens by their voices.

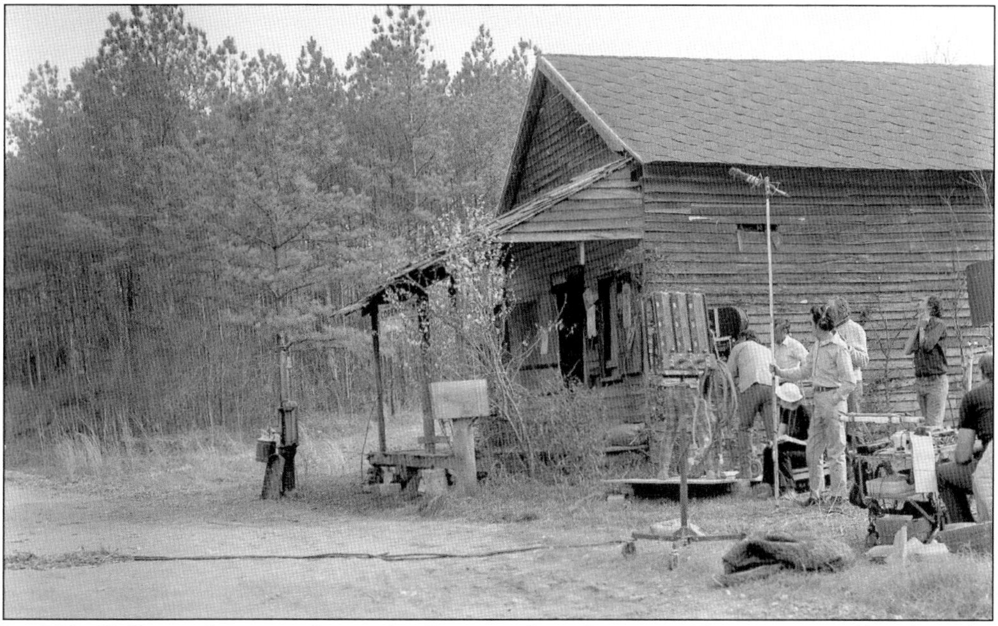

In 1970, Hollywood came to Itawamba County when the movie *Tomorrow* was filmed on the farm of Chester Russell in the Oakland community. A store formerly operated by Jimmy and Ida Chamblee was used in several scenes of the movie, clearly identified as "Chamblee's Store." The store dated to about 1900, and some of the merchandise remained on the shelves from the time of its closure in 1947 and was used in the movie shots. (Courtesy of the Delmus Harden family.)

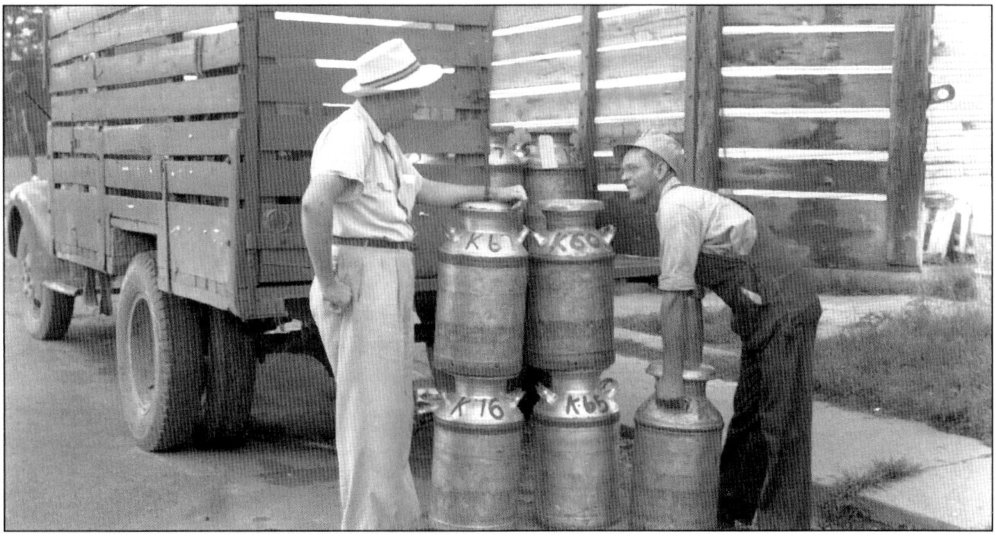

Perdie Montgomery ran a milk route covering northeast Itawamba County in the early 1950s. Hamp Kennedy ran a similar route around Tremont. Then, the two met in Fulton where they transferred Montgomery's load into Kennedy's truck for delivery to the Carnation Milk Plant in Tupelo. County agent H.L. Holland appears to be giving Montgomery some advice as he loads his tanks into the back of the milk truck. (Courtesy of the Delmus Harden family.)

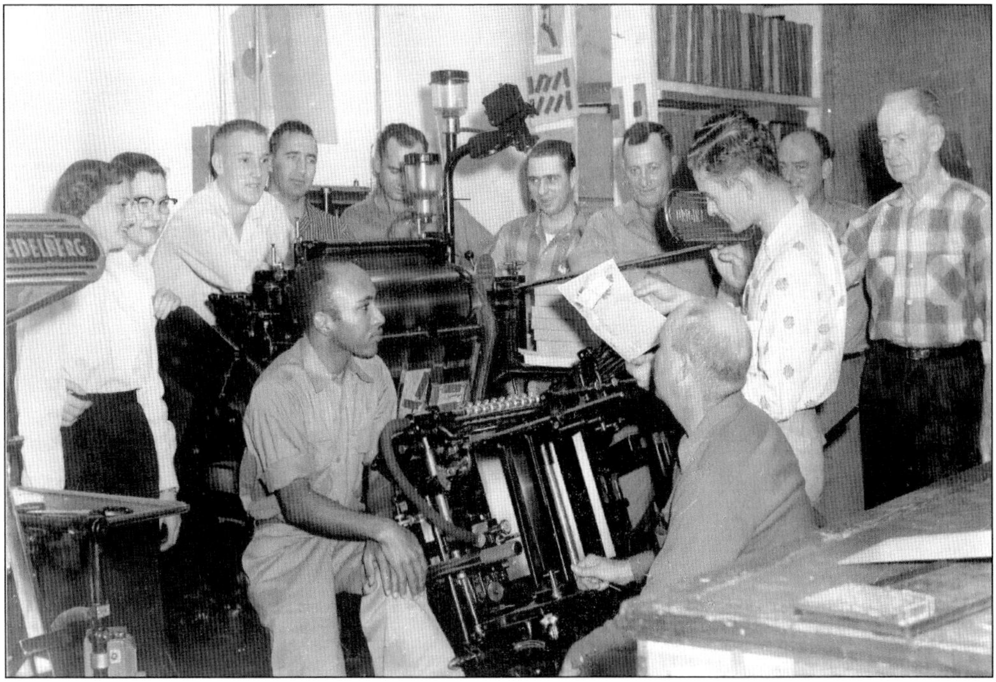

Owned by Delmus Harden from 1945 until his death in 1977, the *Itawamba County Times* was known as the "only newspaper in the world that cares anything about Itawamba County." In 1954, employees gather around the press; seated are Fred Stone, left, and publisher Delmus Harden. Standing, from left to right, are Ruth Howell, Rubye Harden, Bill Miles, Doyle Chandler, Buddy Smith, Don Wallace, Flois Tiddle, Caydo Smith, Earnest Weaver, and Bobby Conn. (Courtesy of Ruth Howell.)

Two

The Muddy Divide

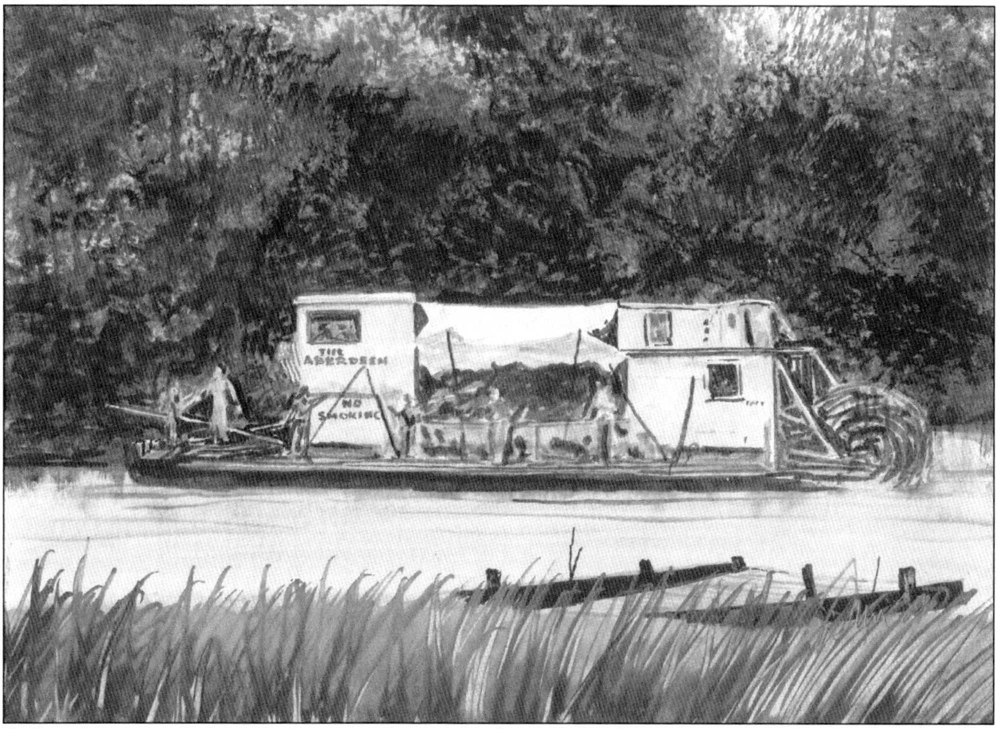

Before there was a county, there was a river. Shallow water, leaning trees, and other obstructions in the Tombigbee River prevented boats from reaching Fulton on a regular, recurring basis. The steamboat *Lillie Lou*, in 1887, was the last boat to reach Fulton until 1909 when a gasoline freight boat named *Aberdeen* (pictured), pulling the barge *Itawamba*, reached the iron bridge at Fulton, where over 1,000 people were waiting to greet it. (Illustration by Arni Anderson.)

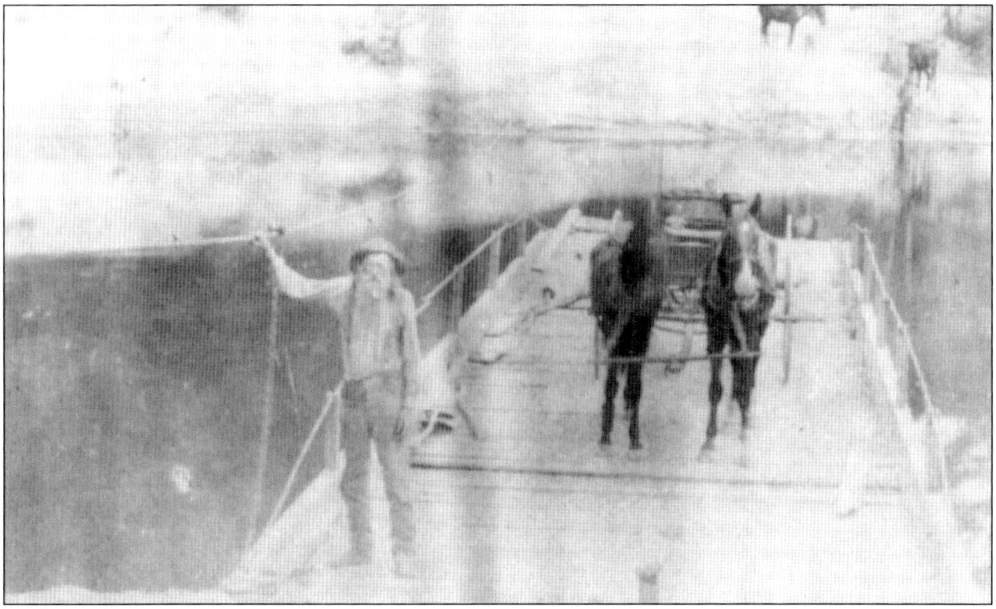

Ferries were the primary means of crossing the Tombigbee River until bridges were built in the early 1900s for crossings at Barr's Ferry, Ironwood Bluff, Bean's Ferry, Fulton, and Walker's Levee. The above ferryboat, with operator Billy Walton, was at the crossing of the Fulton-Pontotoc Road. Using a rope and pulley system, the ferry was pulled across the river by hand. Itawambians crossed for free, but outsiders had to pay a few cents.

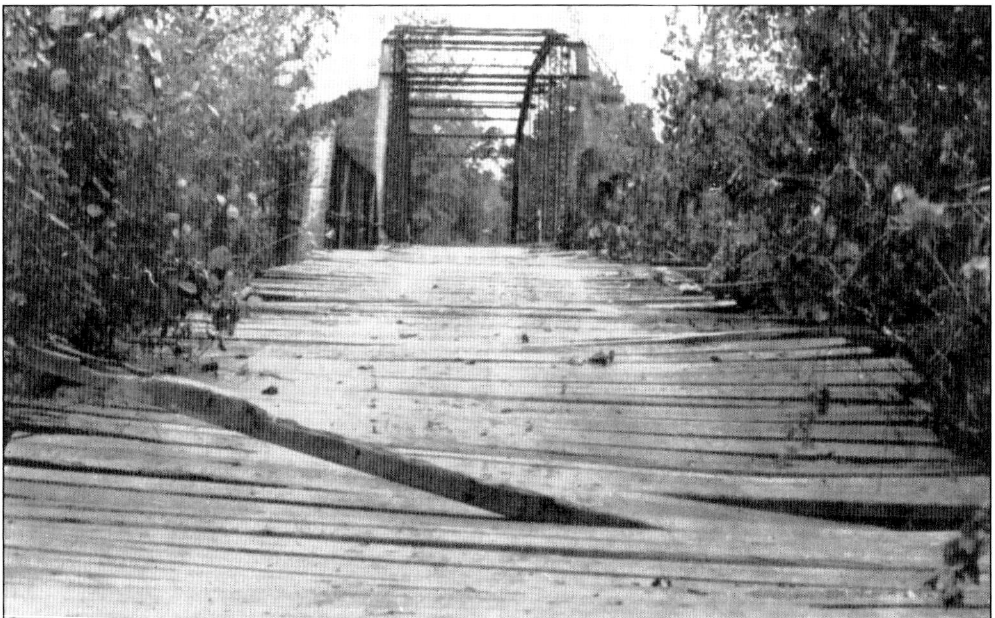

The first bridge to span the Tombigbee River in Itawamba County was finished in April 1907, crossing just above today's Peppertown community and becoming part of the Fulton-Pontotoc Road, later Bankhead Highway in 1921. The iron structure was replaced by a concrete bridge south of the old location when Bankhead Highway was widened in the 1930s. Nothing remains of the iron bridge today. (Courtesy of the Delmus Harden family.)

On October 1, 1921, the Itawamba County Board of Supervisors authorized the building of an iron bridge over the Tombigbee River at Ironwood Bluff, a crossing previously served by ferry. The Vincennes Bridge Company was the lowest bidder at $11,640. The bridge gave farmers in southeastern Itawamba County an easier route to the markets and railroad in Tupelo. (Courtesy of Library of Congress, Prints & Photographs Division, HAER MISS,44-COLUM.V,3-3.)

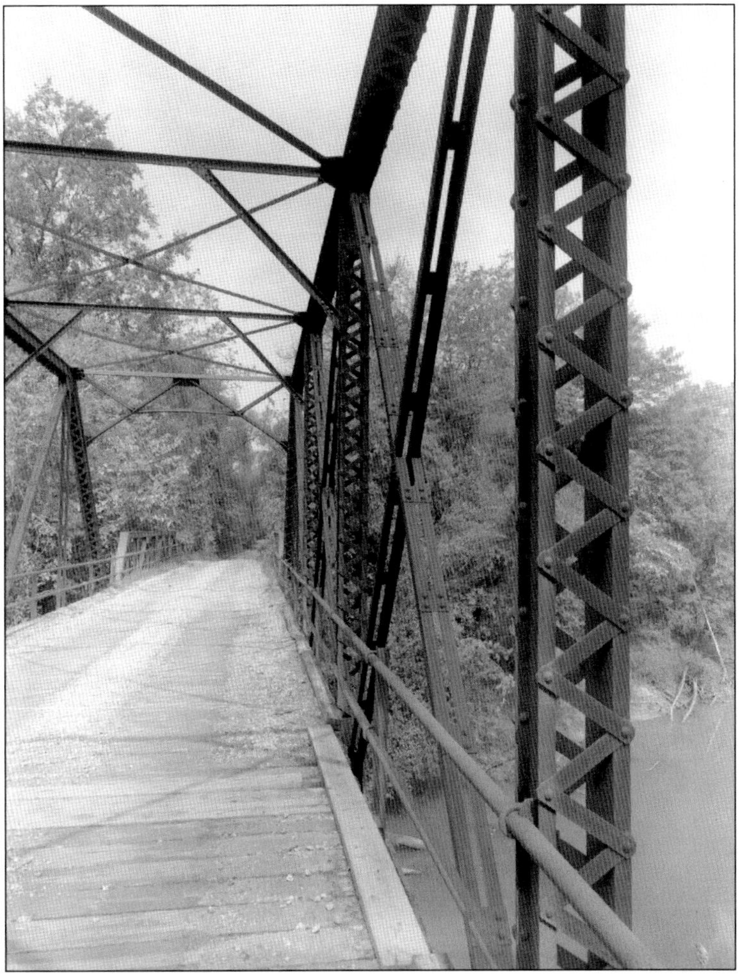

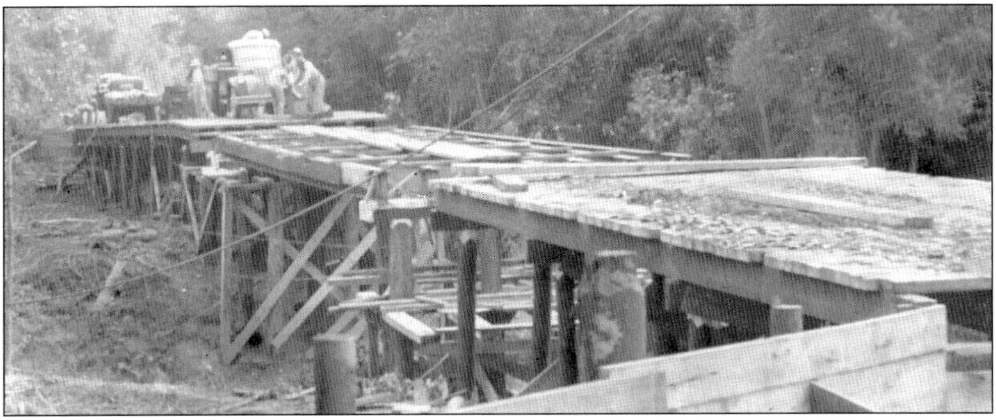

The Bean's Ferry bridge (pictured) was built in 1948. It replaced a 1923 bridge and lasted until construction of the Tennessee-Tombigbee Waterway, although crossing the bridge in its final years was not for the weak-hearted. Earlier, George Bean operated a ferry at the river crossing, hence the name "Bean's Ferry" for this location even today. At least four ferries operated on the old Tombigbee River in Itawamba County. (Courtesy of the Delmus Harden family.)

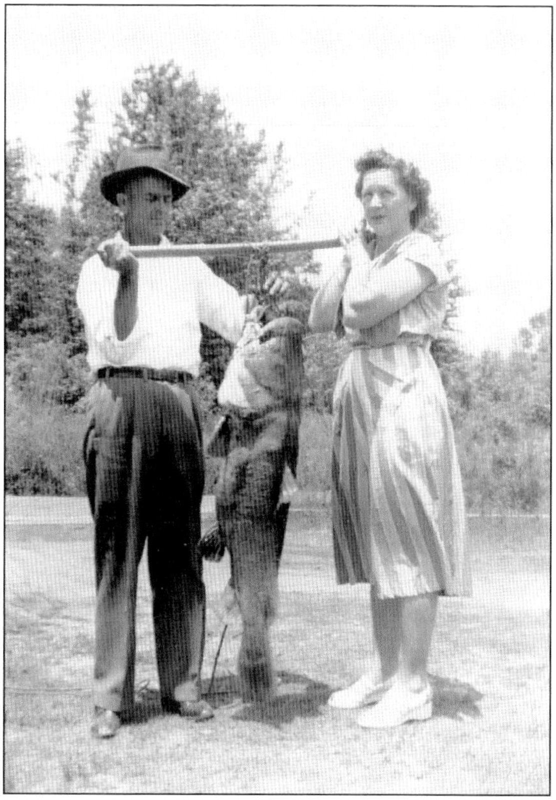

Up and down the county, the Tombigbee River gave sustenance to many Itawamba families. Fish from its waters provided countless meals to the tables of Itawambians like Fessie and Beck Pennington of the Peaceful Valley community. Flathead catfish were the monsters of the river and the goal of every fisherman, but drum and buffalo were other fish that found their way to the dinner table.

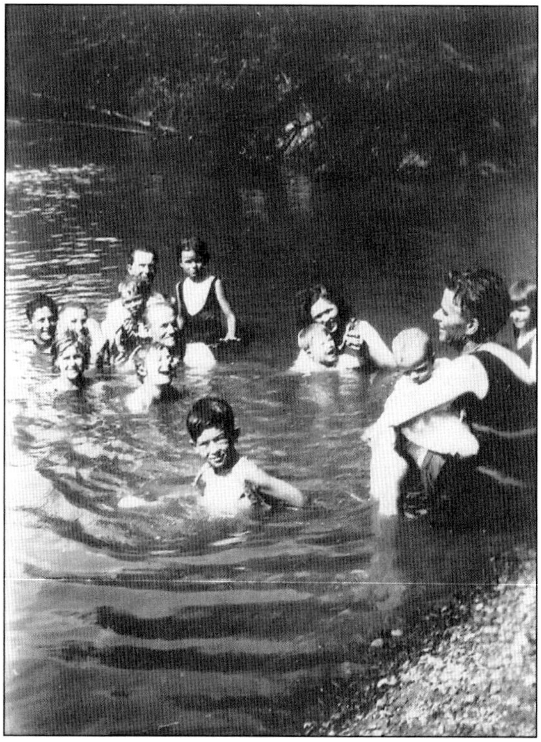

In addition to transportation and food, the Tombigbee River provided opportunities for recreation. The cool, shaded bottom lands of the river were a destination spot for picnics and celebrations during dry summer months. Many an afternoon's delight was enjoyed in a shallow swimming hole or sandbar. The family of Joseph and Augusta Sloan Armstrong, pictured here, enjoyed cooling off in the river. (Courtesy of Linda Brown Pearce.)

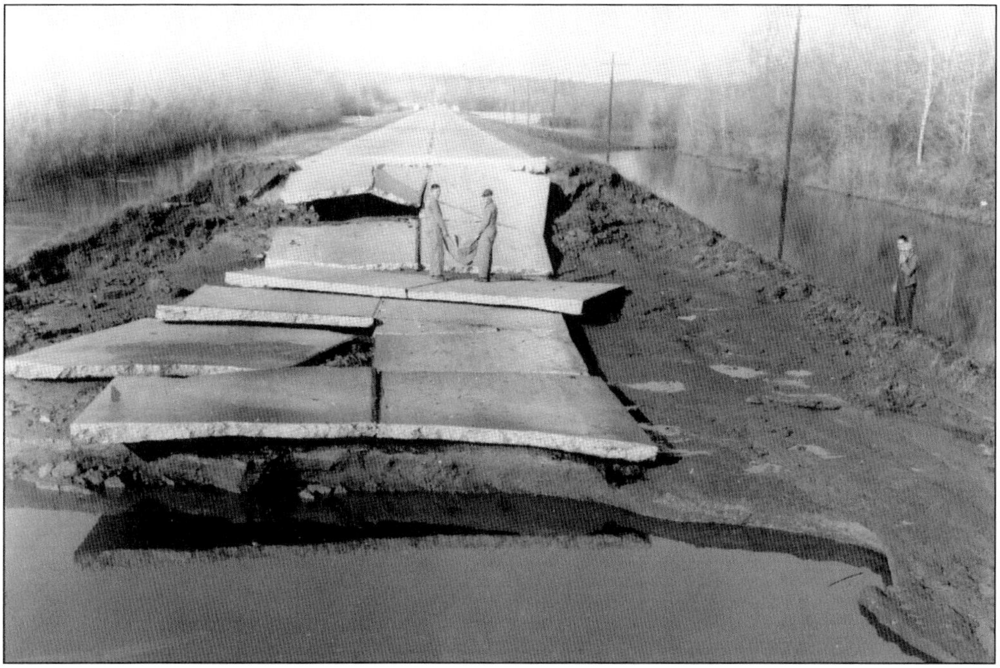

With the Tombigbee River splitting the county in half, Itawambians have particularly been affected by flooding. During record-breaking floods in March 1955, Highway 78 was literally broken to pieces between Fulton and Peppertown. In this photograph, H.D. McGee and Charles Grimes carry the mail sack that was delivered to them at the washout by boat from the Fulton side of the river. (Courtesy of the Delmus Harden family.)

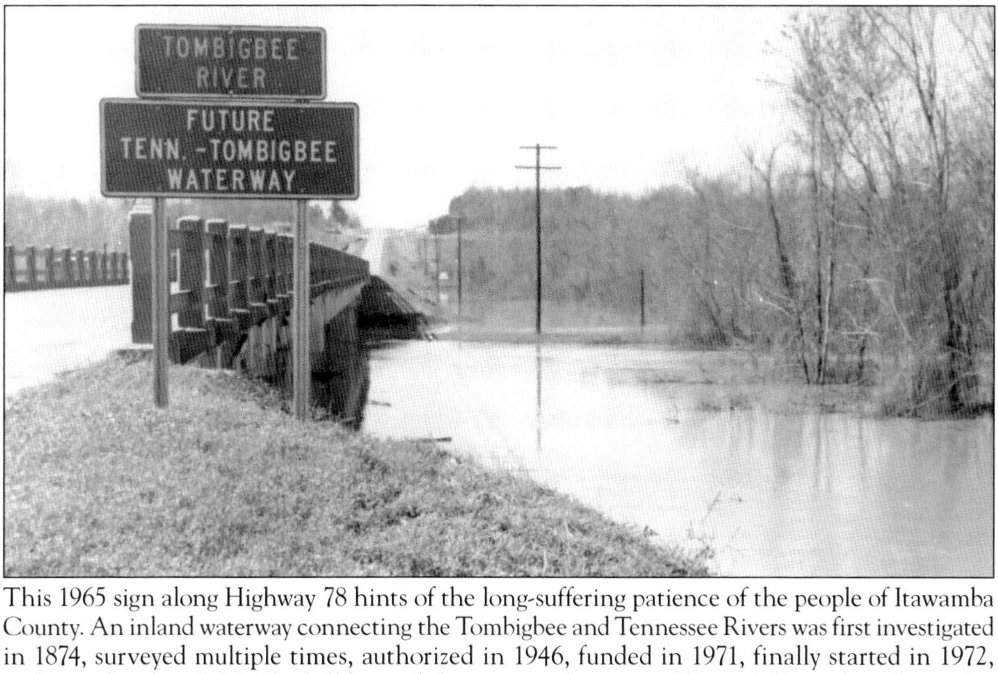

This 1965 sign along Highway 78 hints of the long-suffering patience of the people of Itawamba County. An inland waterway connecting the Tombigbee and Tennessee Rivers was first investigated in 1874, surveyed multiple times, authorized in 1946, funded in 1971, finally started in 1972, and completed in 1984. The hill beyond the sign was known as "River Hill," and just beyond is Peppertown. (Courtesy of the Delmus Harden family.)

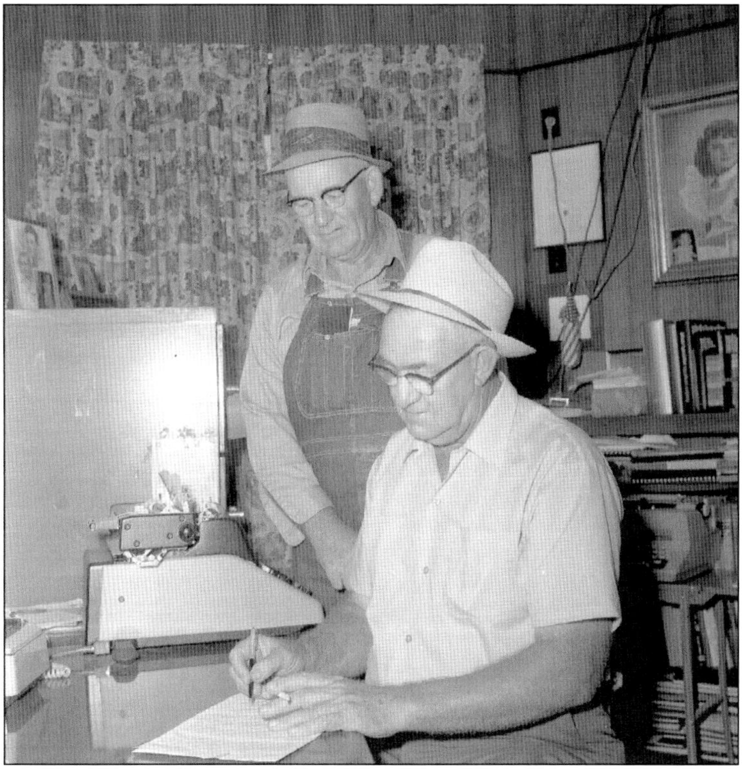

Joe Wheeler Brown (seated) and Oscar Dulaney sign the necessary paperwork on September 24, 1970, that assigned a right-of-way to the US Army Corps of Engineers in anticipation of work to be done in the building of the Tennessee-Tombigbee Waterway. Both men owned land just above the present-day Midway Marina. (Courtesy of the Delmus Harden family.)

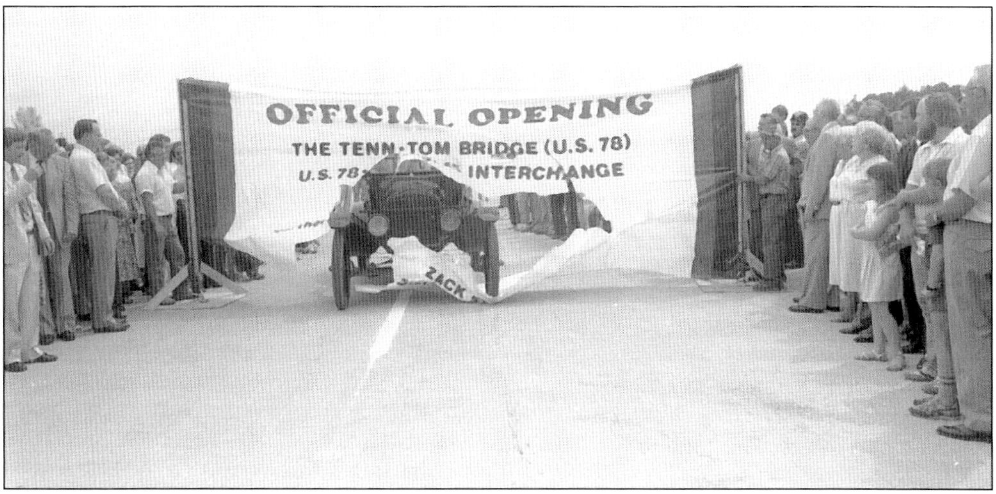

On May 28, 1984, local educator and historian T.L. Burch, in his antique Model T car, drives highway commissioner Zack Stewart through the opening banner across the new Highway 78 Bridge over the Tennessee-Tombigbee Waterway. Hundreds of Itawambians turned out for the historical event. The old highway across the Tombigbee River was closed forever on June 10, 1984, and on January 10, 1985, the first barge came through the county. (Courtesy of the Delmus Harden family.)

Three

Musical Heritage

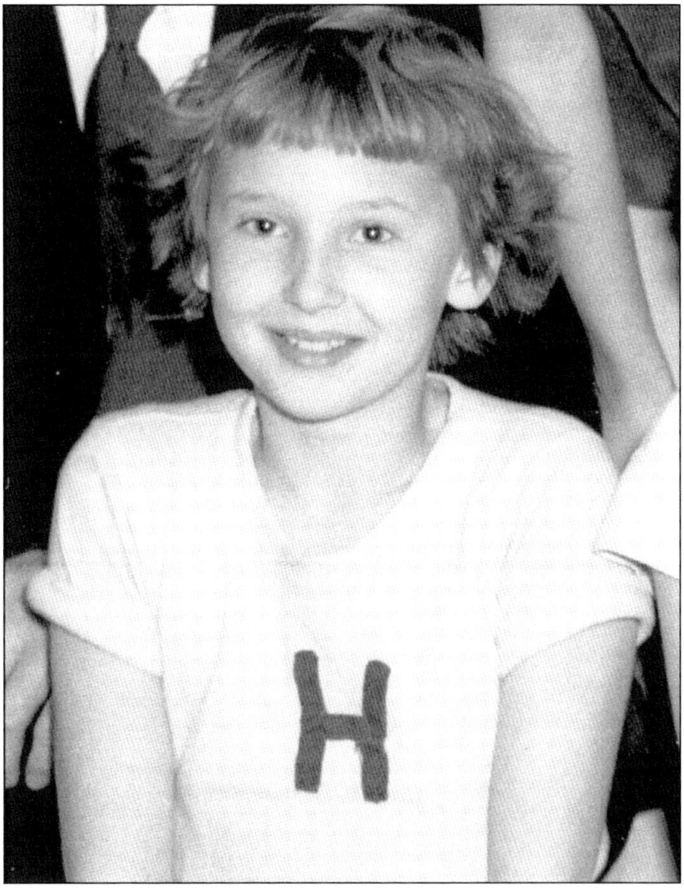

Itawamba County enjoys a wonderfully rich musical heritage. Who could believe that Tammy Wynette (the "First Lady of Country Music"), Elvis Presley (the "King of Rock and Roll"), and Jimmie Lunceford (the "King of Syncopation") all share common roots in the rolling hills and red-dirt farms of Itawamba County? Little Wynette Pugh (Tammy Wynette), shown here as a fourth grader, received her early education at Hopewell School near Tremont.

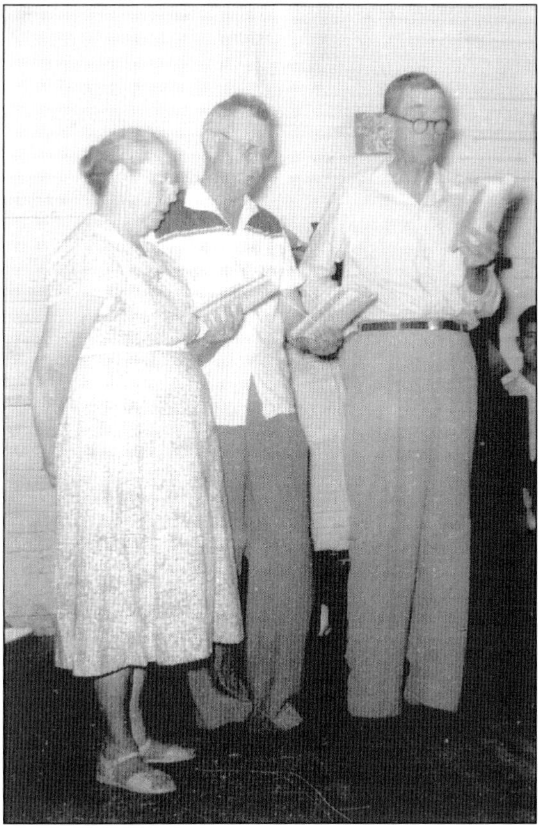

Nora Thornton Johnson and her brothers, Ira Green Thornton and John Thomas Thornton, lead the singing at Sandy Springs Baptist Church. Nora was said to have had a beautiful alto voice. Their first cousin, Prof. Allie Patton Thornton, was a popular instructor at singing schools held throughout the county and was in demand at all-day singings as well. (Courtesy of Glenda Johnson.)

The Sloan family, of the Peaceful Valley community in Itawamba County, had a love for music that is carried on today by descendants at holidays, family reunions, and other gatherings. Pictured from left to right are Dorothy Sloan, Afton Sloan, Bill Newton (brother-in-law), and Grace Sloan. These ladies were daughters of Gainey Sloan and Dora Ridings; Bill was the son of Elonzo Newton and Maude Bourland. (Courtesy of Gail Newton Carter.)

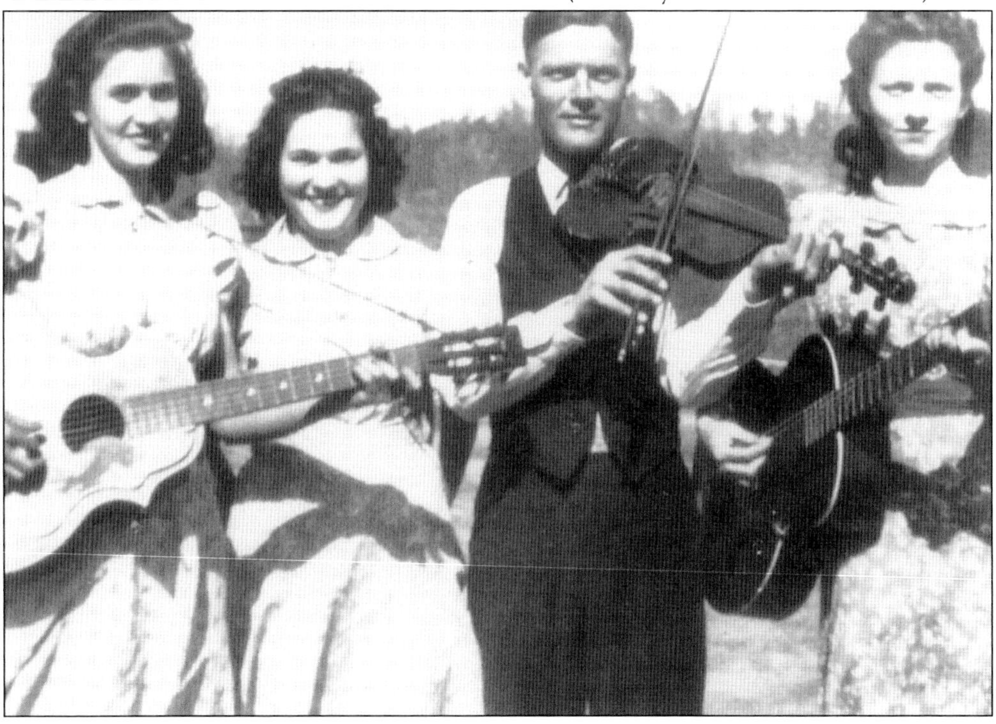

The Beam family of Salem and Fairview communities loved their music. After a long week of farming or sawmilling, they would gather at someone's home, push the furniture back against the wall, and start playing their music. Seated with their instruments are Calvin Beam (left) and Carl (Calvin's nephew), while behind them are Ruble (Calvin's son) and Gurney (Calvin's nephew). (Courtesy of Hazel Beam Harnage and Dorothy Beam Moody.)

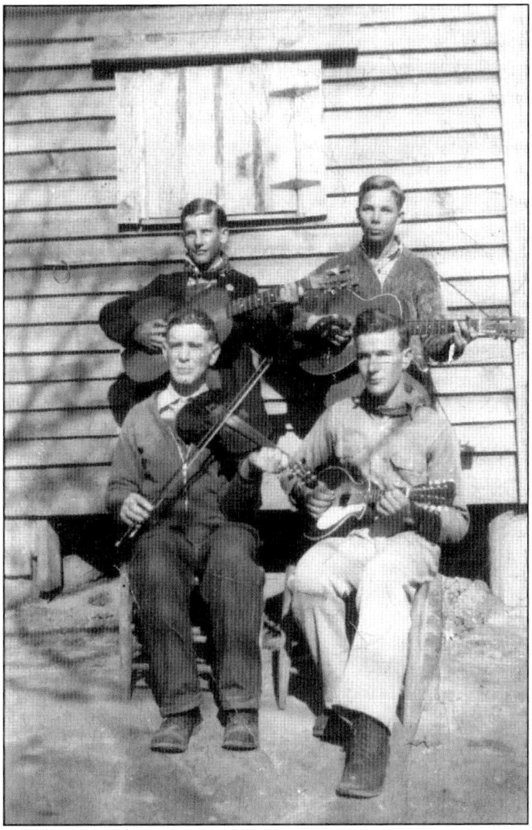

From left to right in this 1935 photograph, brothers Vonnie, Donnis, Adron, and Earnest Johnson started playing music at a young age at their home near Sandy Springs (not pictured is their brother Julius). Nearly every Johnson gathering included music with a little bit of buck dancing thrown in. These brothers were the sons of Fisher and Nora Thornton Johnson. (Courtesy of Glenda Johnson.)

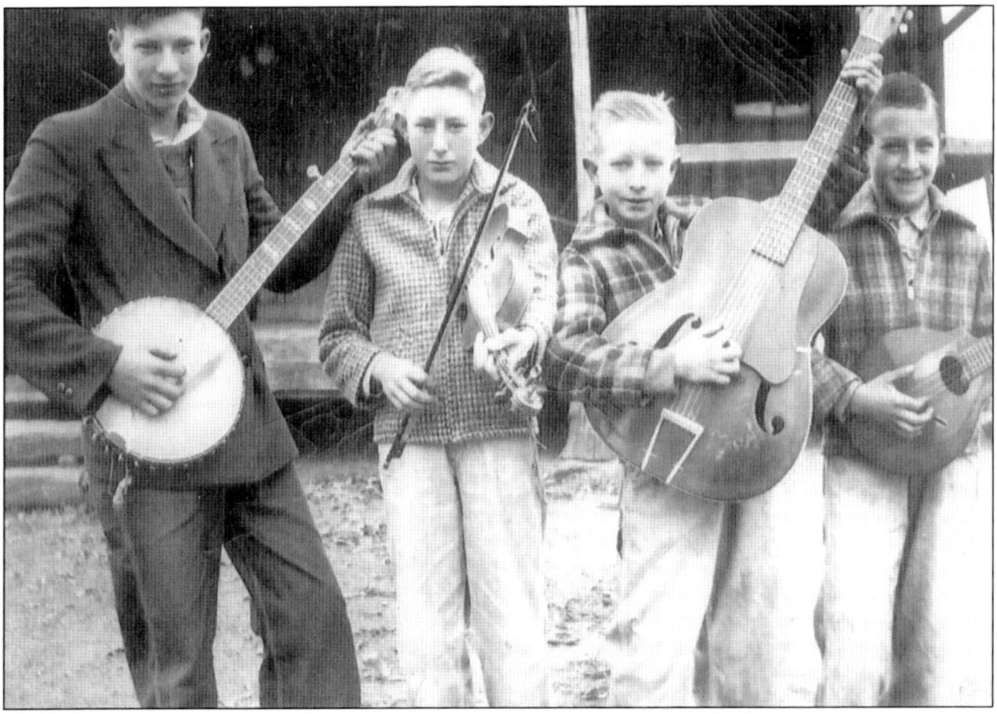

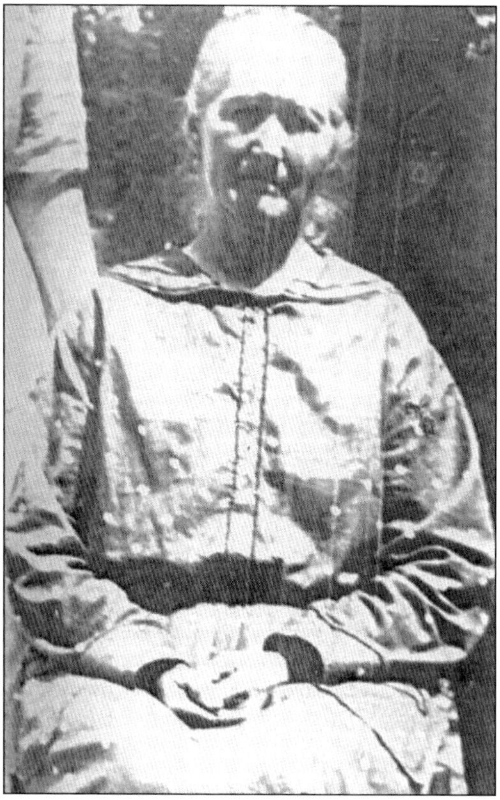

Rosella Presley was born around 1862 in Itawamba County to Martha Jane Wesson and Dunnan Presley Jr. Rosella never married, had nine children, and supported her family by sharecropping. Her grandson Vernon Presley, also born in Itawamba County, was the father of the King of Rock and Roll, Elvis Presley. Rosella died in 1924 and was buried in Ridge Cemetery, not far from where she last lived. (Courtesy of Julian Riley.)

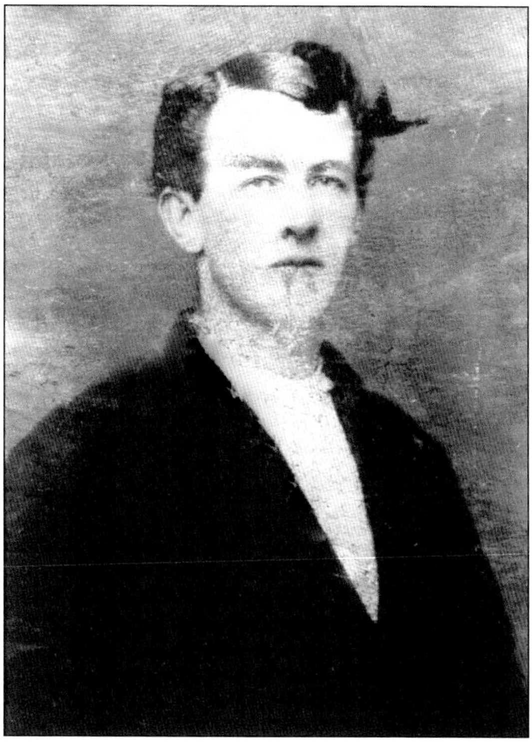

Fiddle-playing John Wallace is said to have been the father of Jesse Dee Presley, and the great-grandfather of Elvis. His relationship with Rosella Presley has been documented by oral family history and acknowledged by Jesse's son, Vernon Presley, in conversations with Wallace family members. (Courtesy of Mackey Hargett.)

Affectionately called "Dodger" by her grandson, Elvis Presley, Minnie Mae Hood Presley was born in Itawamba County in 1888. She was the daughter of William H. and Minerva Warren Hood, both of whom came from pioneering families of eastern Itawamba County. Minnie married Jesse Dee Presley about 1913. Pictured in 1957 is Minnie, left, with her sister, Minerva, during a family gathering back home in Itawamba County. Minnie lived at Graceland until her death in 1980. (Courtesy of Julian Riley.)

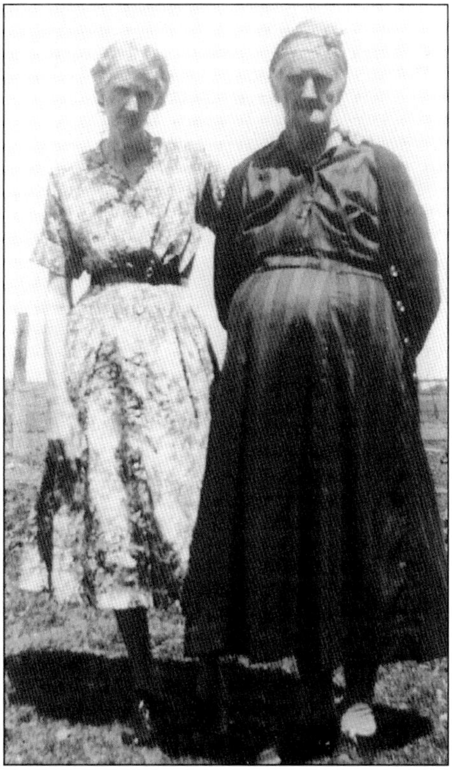

Mackey Hargett, great-grandson of John Wallace and his wife, Elmira McFadden, stands beside the historical marker north of the Mt. Pleasant community in Itawamba County that identifies the birthplace of Vernon Presley, grandson of John Wallace and Rosella Presley. At Graceland, when Mackey was introduced to Elvis by Vernon, Vernon told Elvis that Mackey "was a cousin on the Wallace side of the family."

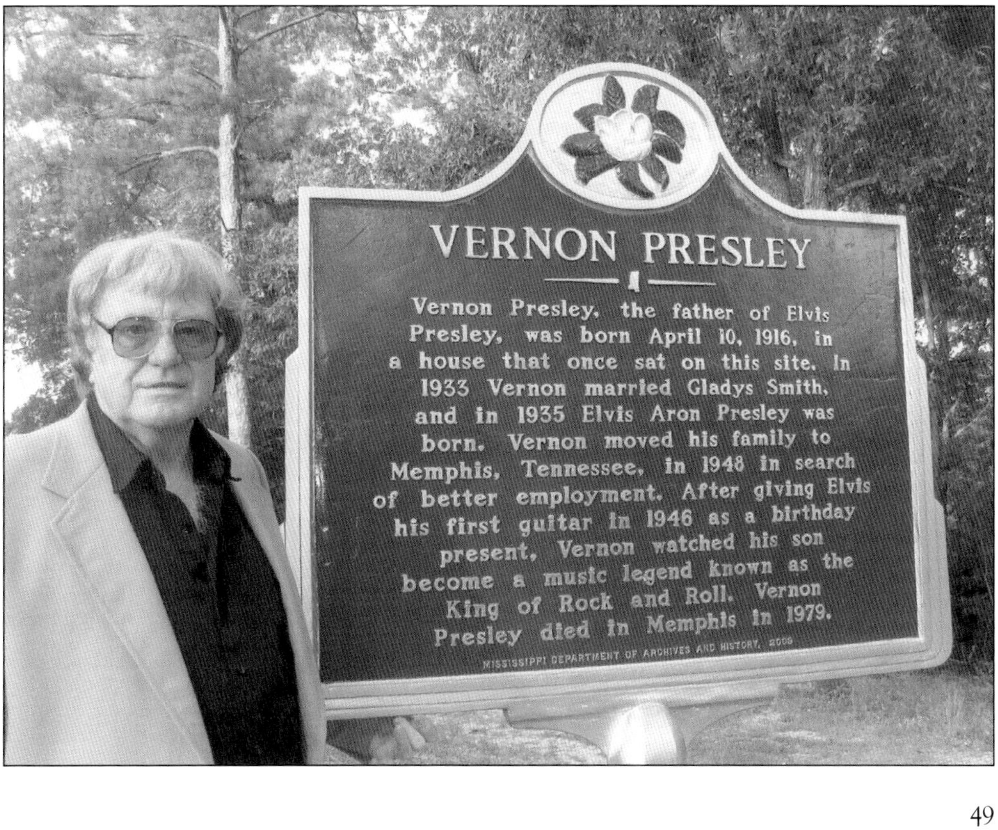

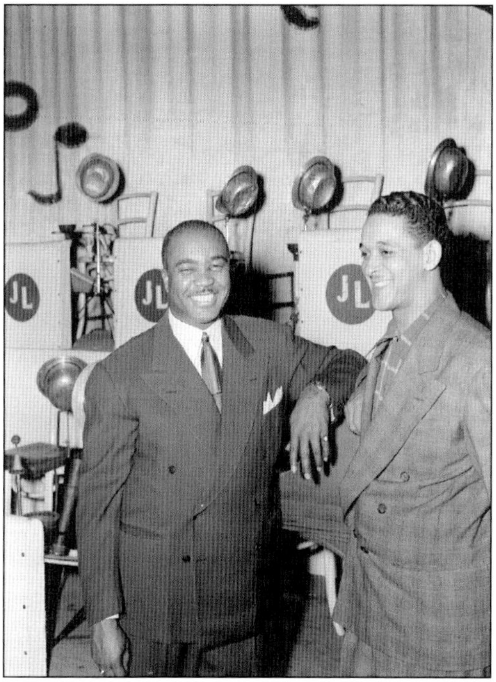

Jazz great James Melvin "Jimmie" Lunceford, left, was born in Itawamba County in 1902, the elder son of James Riley and Idella Shumpert Lunceford. Jimmie's grandparents, Daniel and Gracy Lunceford, are buried in Palmetto Cemetery northeast of Fulton. Although his family later moved to Colorado, Jimmie returned to the South and taught school at Memphis, where he organized a student band called the Chickasaw Syncopators. Jimmie is pictured here with his trombonist Trummy Young. (Courtesy William P. Gottlieb/Ira and Leonore S. Gershwin Fund Collection, Music Division, Library of Congress.)

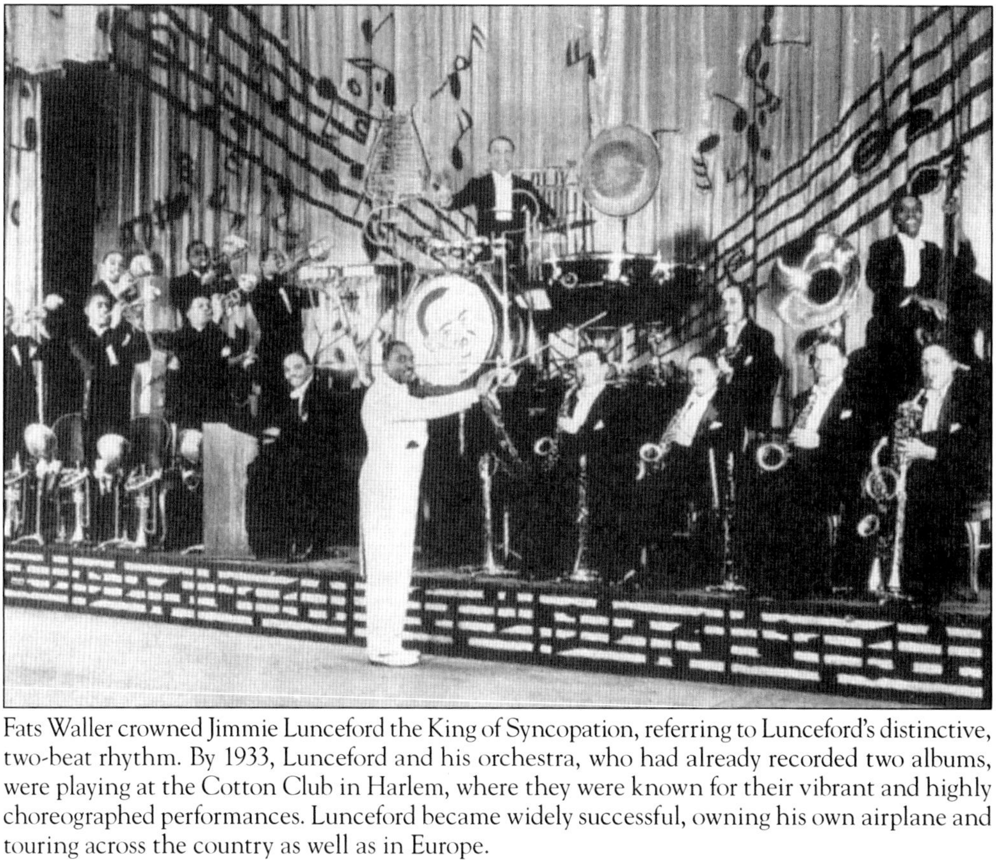

Fats Waller crowned Jimmie Lunceford the King of Syncopation, referring to Lunceford's distinctive, two-beat rhythm. By 1933, Lunceford and his orchestra, who had already recorded two albums, were playing at the Cotton Club in Harlem, where they were known for their vibrant and highly choreographed performances. Lunceford became widely successful, owning his own airplane and touring across the country as well as in Europe.

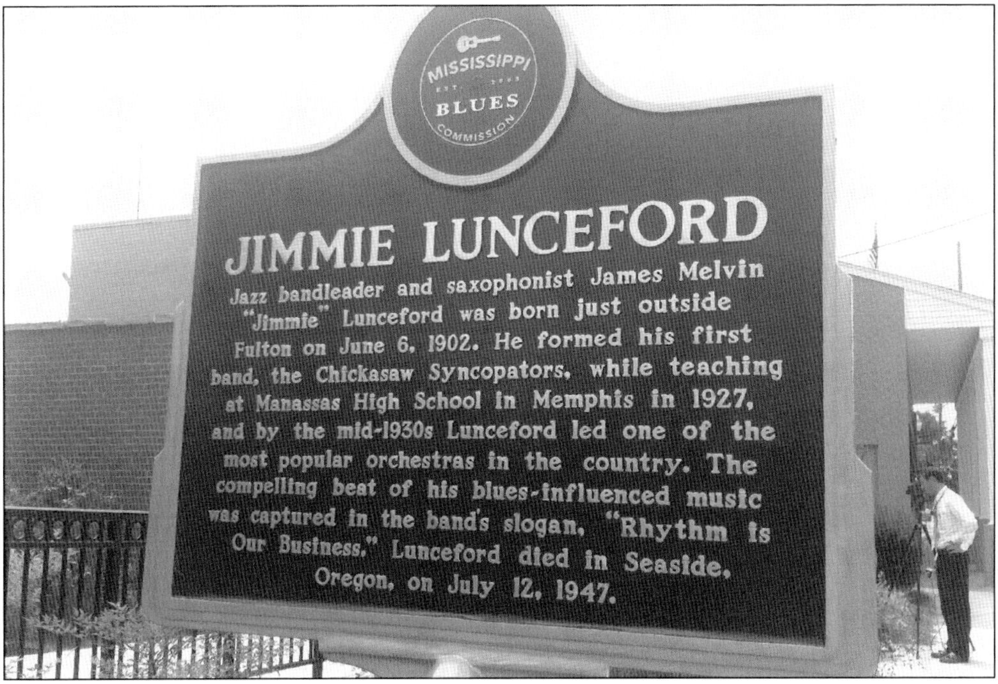

Itawamba County, in 2011, honored Itawamba-born Jimmie Lunceford with a historical marker for his contributions to the nation's musical heritage. More popular than Cab Calloway during his time, Lunceford was both admired and copied by renowned orchestra leader Glenn Miller, who once remarked, "Duke is great, Basie is remarkable, but Lunceford tops them both."

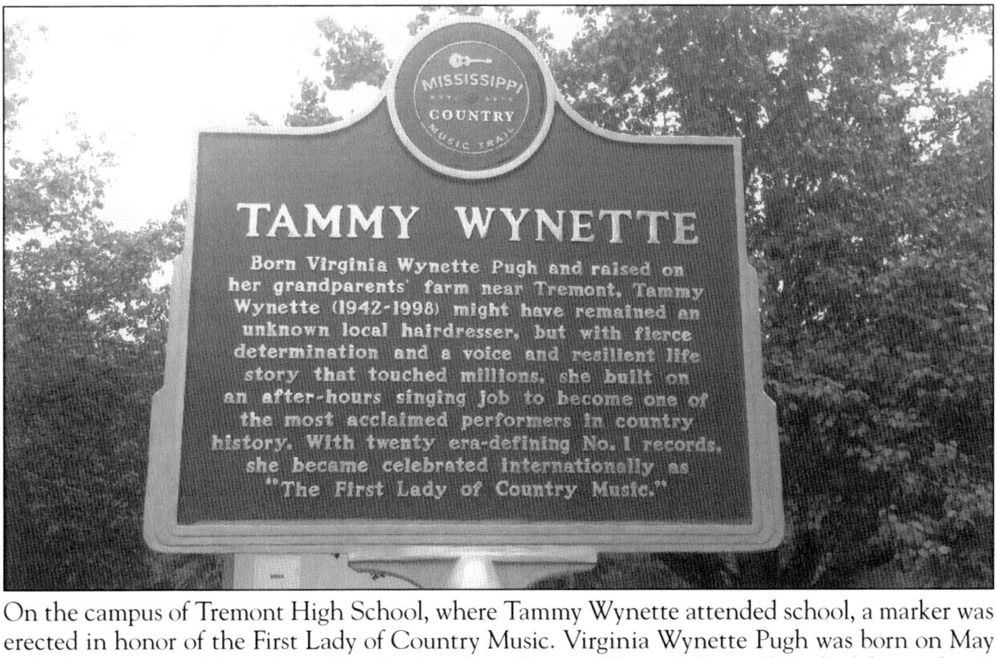

On the campus of Tremont High School, where Tammy Wynette attended school, a marker was erected in honor of the First Lady of Country Music. Virginia Wynette Pugh was born on May 5, 1942, to Mildred Russell and her husband, William Hollice Pugh. Hollice died from a brain tumor shortly after Wynette's birth, but he passed down his musical talent and love of music to his daughter.

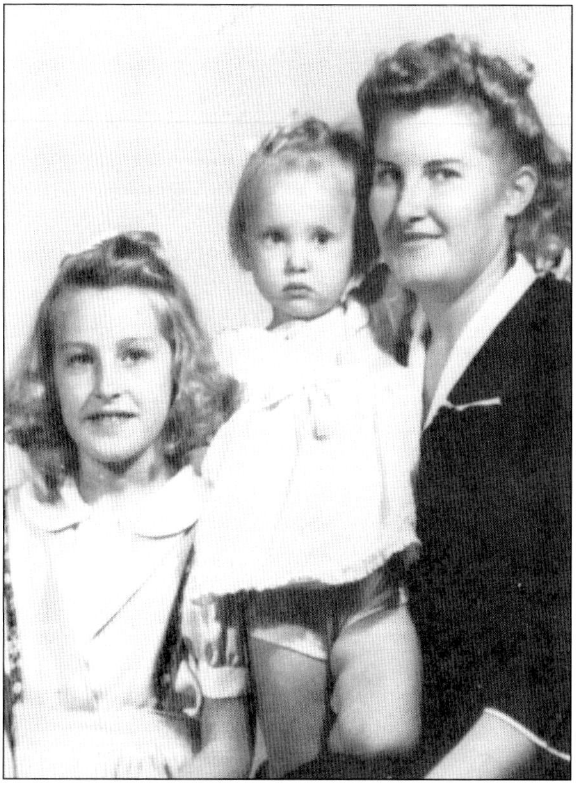

Tammy's grandparents, Chester and Flora Allen Russell, practically raised her. Their younger daughter, Carolyn, was only five years older than Tammy and was more like a sister than an aunt to her. The Russells lived in the Oakland community in eastern Itawamba County, where Chester was a farmer and sawmiller and served on the board of trustees for Itawamba Junior College. His elder daughter, Mildred, the mother of Tammy Wynette, is pictured with her only daughter, center, and her own much younger sister, Carolyn.

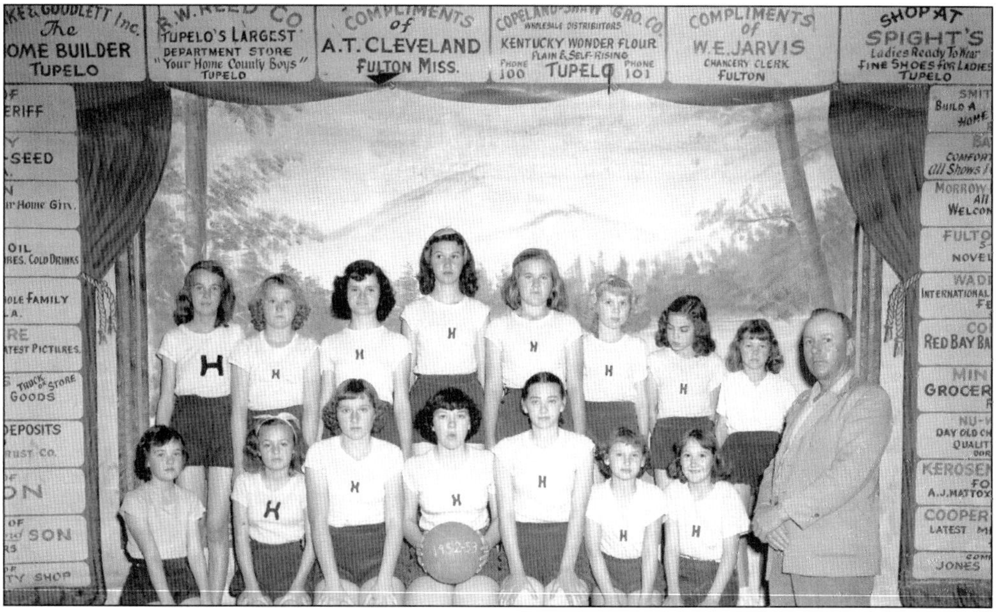

Hopewell School in the Bounds Crossroads community was where Wynette Pugh received her early schooling. Principal T.L. Burch coached the girls' basketball team in 1953; Tammy is in the front row, second player from the right. Tammy could sing *and* dribble, once saying that she loved basketball more than anything except singing. In high school, she received All-State honors on Tremont's basketball team. (Courtesy of the Delmus Harden family.)

Wynette Pugh and Doug Horne were voted Most Popular at Tremont High School in 1959. Tammy, who played flute in the school band, was named Miss Tremont High as well. Tammy attended church at both Oak Grove Church of God, where she played the guitar and helped lead the singing, and Providence Baptist Church, where she started playing the piano for the congregation at age eight. (Courtesy of Doug Horne.)

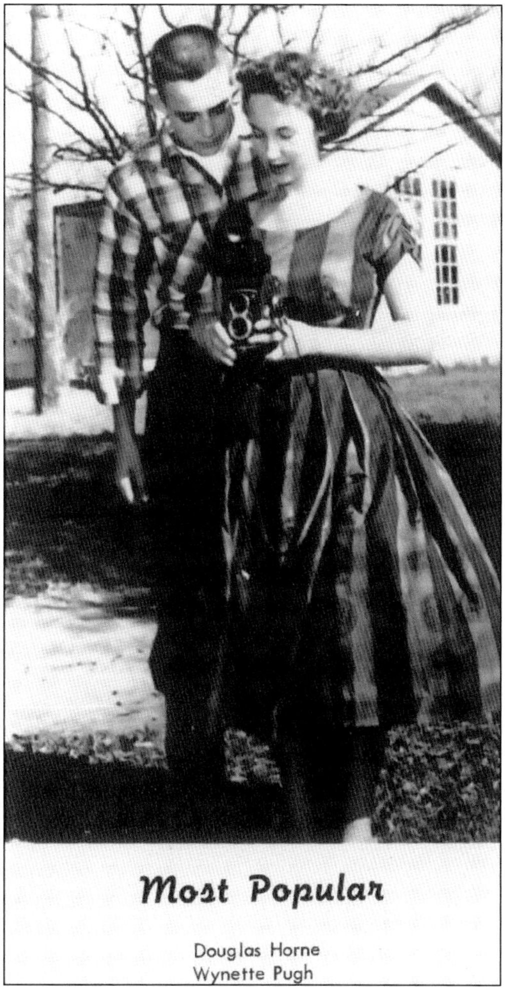

This was the scene for many years in the courtroom on the second floor of Itawamba County's courthouse: regular Sunday afternoon nondenominational "singing conventions." In this 1951 photograph, sacred harp singing is being conducted in the court room by a song leader inside the traditional circle or square. Singers are reading shaped notes out of a special hymnal book. (Courtesy of the Delmus Harden family.)

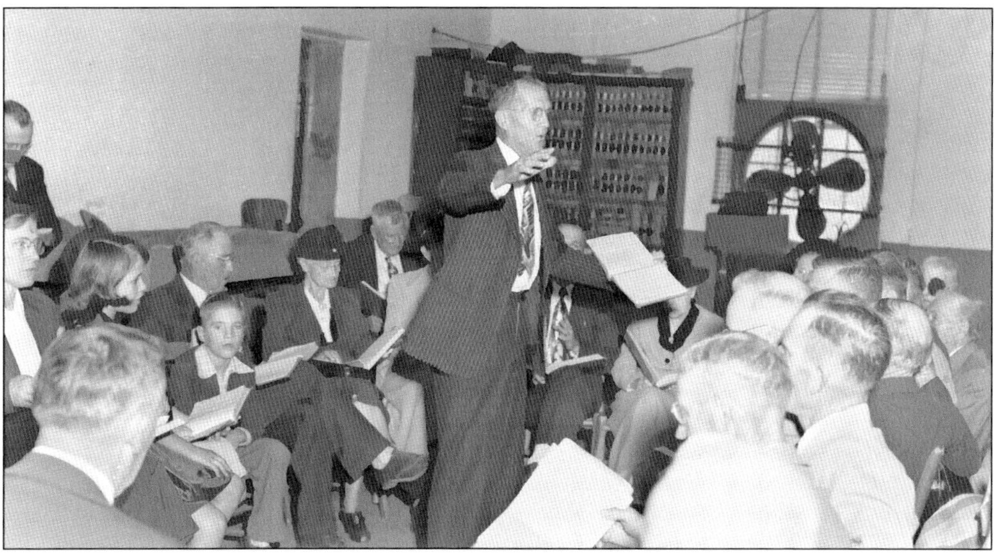

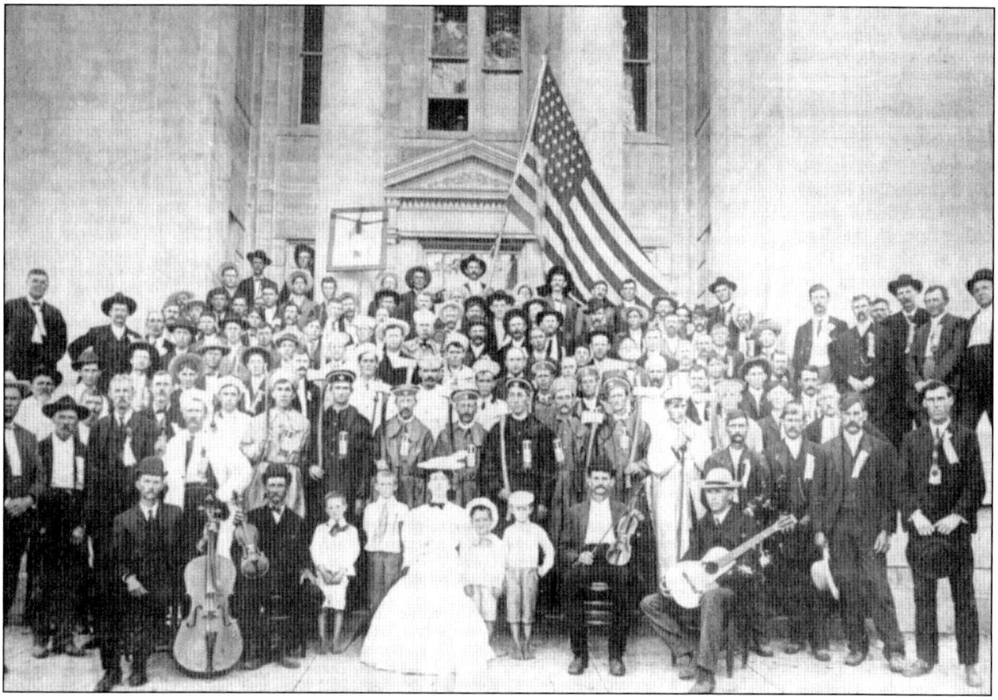

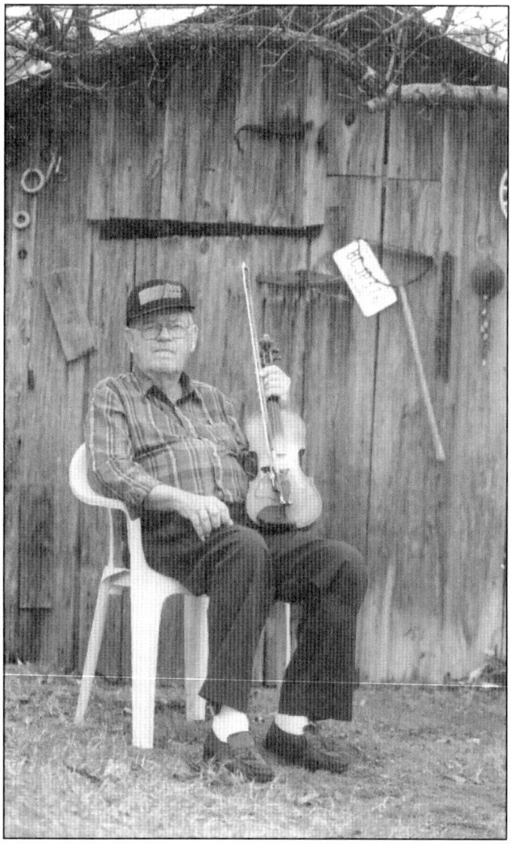

The Cockrells were one of the more well-known musical families of Itawamba County. Above, in this 1890s photograph taken at a Woodmen of the World gathering in front of the Lee County Courthouse, members of the Cockrell String Band sit in the front row with their instruments. From left to right, they are William Cockrell, Elijah Cockrell, Jordan Cockrell, and ? Cummings. Jordan left his home in Itawamba County in 1904 and traveled to the world's fair in St. Louis, where he entered and won the World's Fiddling Championship. Jordan's younger sister, Nancy Cockrell Thornberry, played the dulcimer in addition to a fiddle.
The Cockrells descend from Elum and Caroline Devaughn Cockrell, who moved to Itawamba County in the 1870s. At left, another Cockrell descendant, Carl "Skeet" Grissom, has delighted audiences for years with his fiddle. Grissom won the fiddle contest at the 1945 Fair and Dairy Show in Tupelo, the same year that Elvis Presley took fifth place in a talent contest for singing "Old Shep." (Left, courtesy of Carl Grissom.)

Four

FACES FROM THE PAST

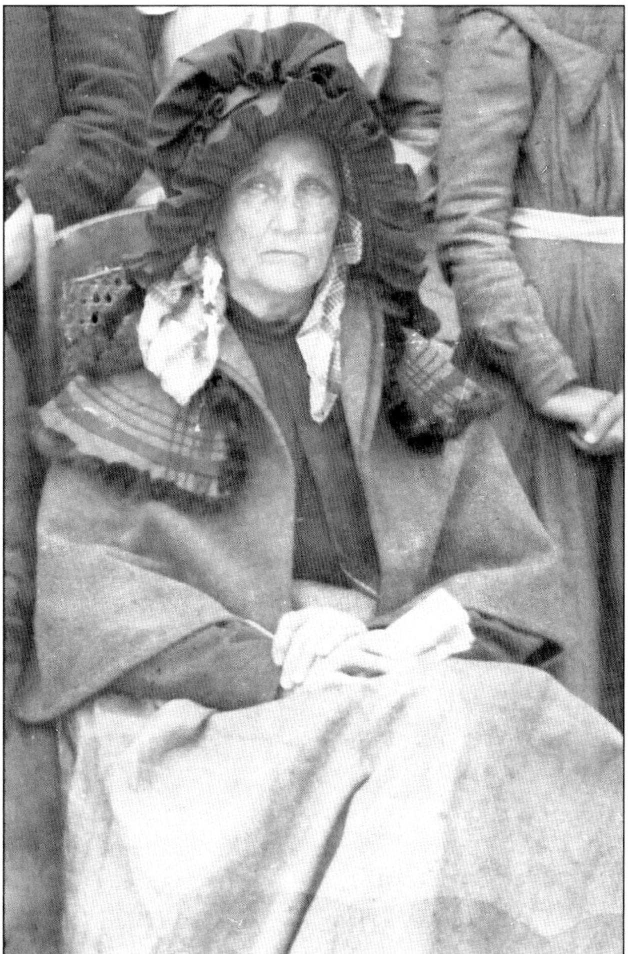

Harriet Caroline Pearce Johnson was the only one of her Pearce family to travel to Itawamba County from her native North Carolina. She and her husband, Stephen, arrived around 1855 from Wake County and settled near Gum Creek, where many of their descendants live today. Although the Johnsons attended Mt. Pisgah Baptist Church initially, at least three of their sons were charter members of the Pine Grove Church of Christ.

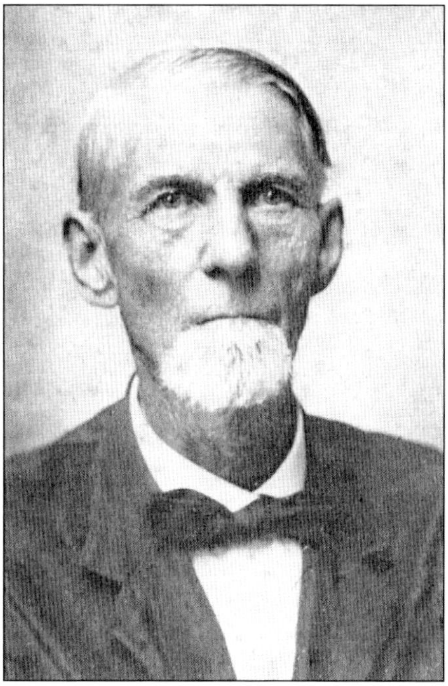

Washington Lafayette Clayton came to Itawamba County with his family in 1840. Clayton, a lawyer and Civil War veteran, wrote newspaper accounts about frontier life in the early days of Itawamba County. These stories were later collected into a book compiled by his great-granddaughter Minrose Gwin. Clayton's narratives serve as a valuable resource for historians. (Courtesy of Minrose Gwin.)

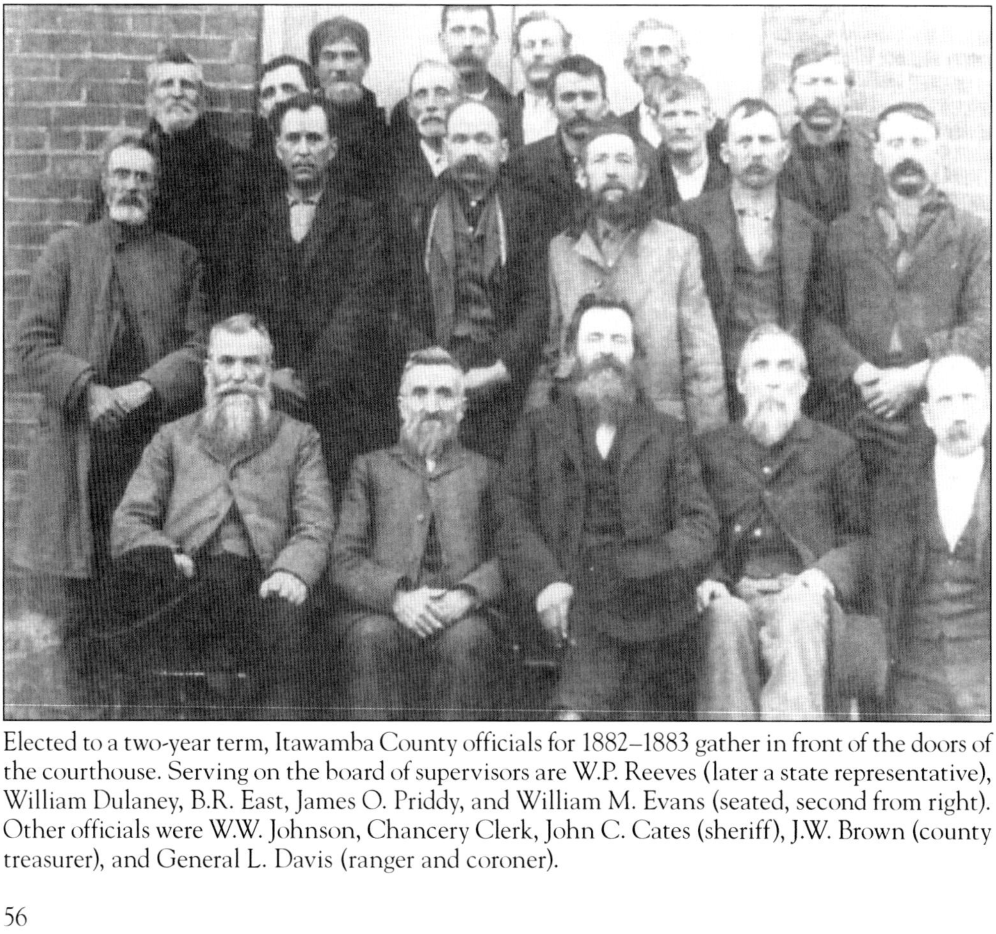

Elected to a two-year term, Itawamba County officials for 1882–1883 gather in front of the doors of the courthouse. Serving on the board of supervisors are W.P. Reeves (later a state representative), William Dulaney, B.R. East, James O. Priddy, and William M. Evans (seated, second from right). Other officials were W.W. Johnson, Chancery Clerk, John C. Cates (sheriff), J.W. Brown (county treasurer), and General L. Davis (ranger and coroner).

Mary Elizabeth Friday, daughter of William H. and Malinda C. Friday, was born in 1858 into four early pioneering families of Itawamba County. Her paternal grandparents were Reuben A. Friday and Sarah Elizabeth Green, while her maternal grandparents were Jacob Floyd Martin and Patience Clayton—all early settlers of the county. Rev. Jacob F. Martin is credited with establishing Mt. Pleasant Methodist Church, serving as its first minister. (Courtesy of Patsy Adcock Kennedy.)

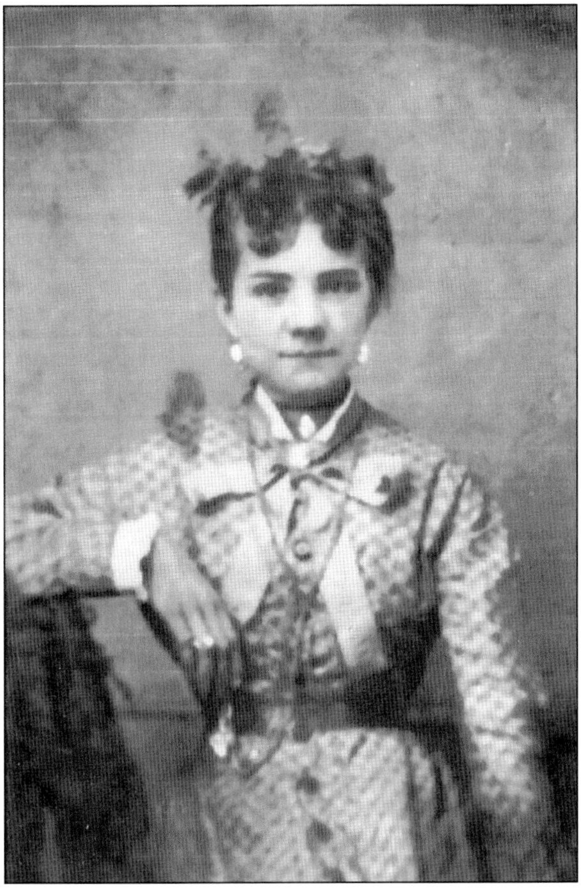

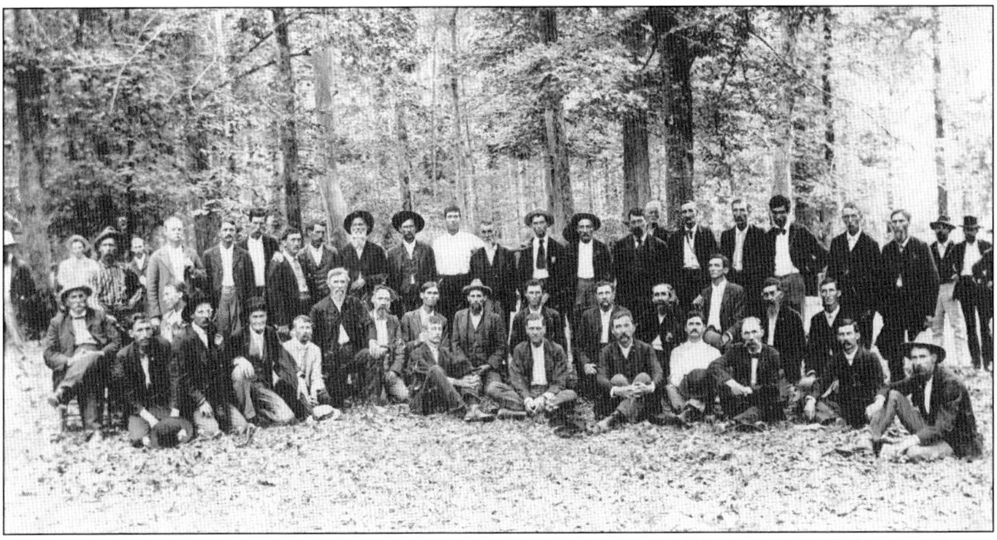

Candidates gather for a group photograph in the late 1800s at one of the many candidate speakings held throughout the county during election year. These events were popular as both political and social occasions and attracted large crowds of people. White Springs Resort in southern Itawamba County was always a popular venue for such activities.

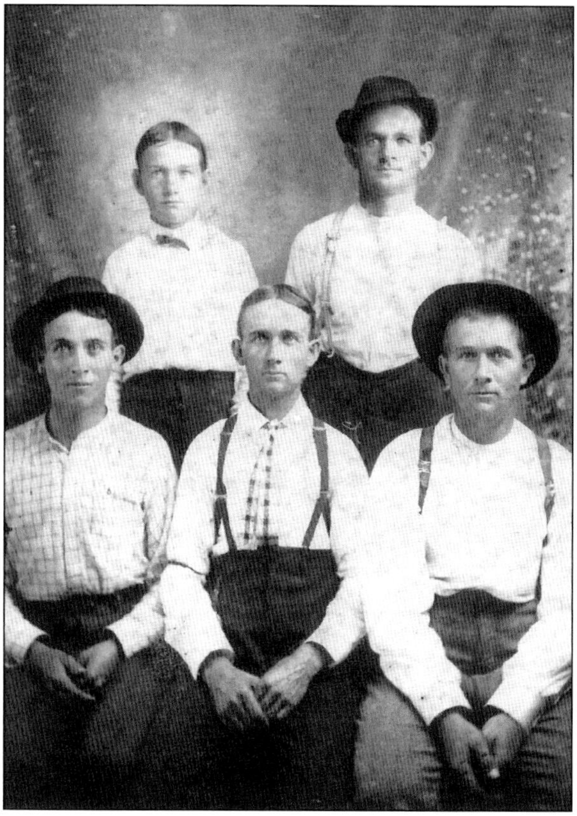

James William Anderson "Billy" Davis and Annaliza Morrow had five sons, photographed here about 1902. From left to right are Leon Burg, Dew Boyd, John Stonewall, Jesse McGee, and James Kelly. The couple also had three daughters—Anna Luna, Lizzie Lee, and Hettie Anderson. Billy, the son of Jesse and Elvira McGee Davis, and Annaliza, the daughter of John and Mary Ford Morrow, were among the founders of Fulton Baptist Church.

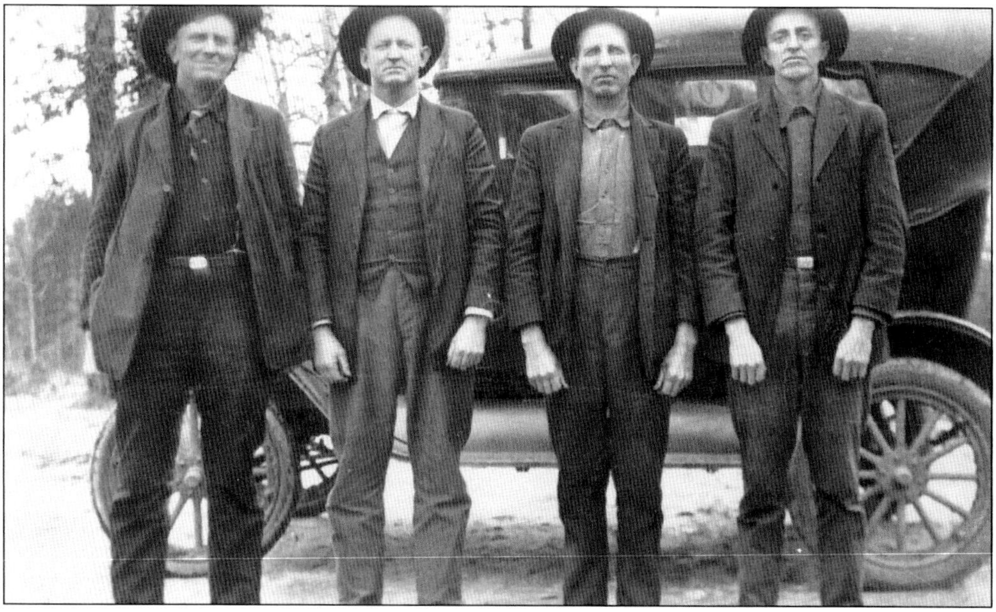

Benjamin Bradford Miles was an Illinois native who came to Itawamba County and, in 1872, married Martha Ann Johnson, daughter of Larkin and Mary West Johnson. Pictured here are the couple's sons, from left to right, William "Will," Larkin Leroy "Lee," Ellie Hugh, and Ira Matison "Mat." Other children not pictured were Jennie, James, Mary, Martha Frances, and Susan Florence.

Thomas L. McFerrin and Inez Marlin were married on February 11, 1900. After being widowed in 1919, Thomas moved from the Mooreville area, where he had a store, to Mantachie with his three small sons, Gordon Marlin, Orville Kermit, and Robert Keyron. He continued in the merchandising business by purchasing an existing store, which later became known as McFerrin Brothers Store.

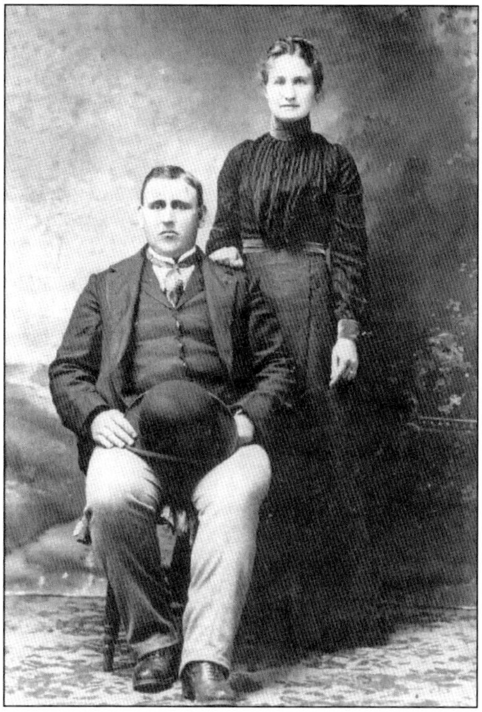

John Dulaney was among a handful of settlers already living in the area that became Itawamba County as found by surveyors in 1835. John and his brothers, Alfred and Gilbert, moved here with Rhoda, their mother. Pictured from left to right are the children of Alfred and Rachel McNiece Dulaney in 1905: (first row) Tom, John (holding cane), Sibbie, Caroline, Mary, and William; (second row) Gilbert, Alfred "Babe," and Joseph (with hat on head). (Courtesy of Virble Booth.)

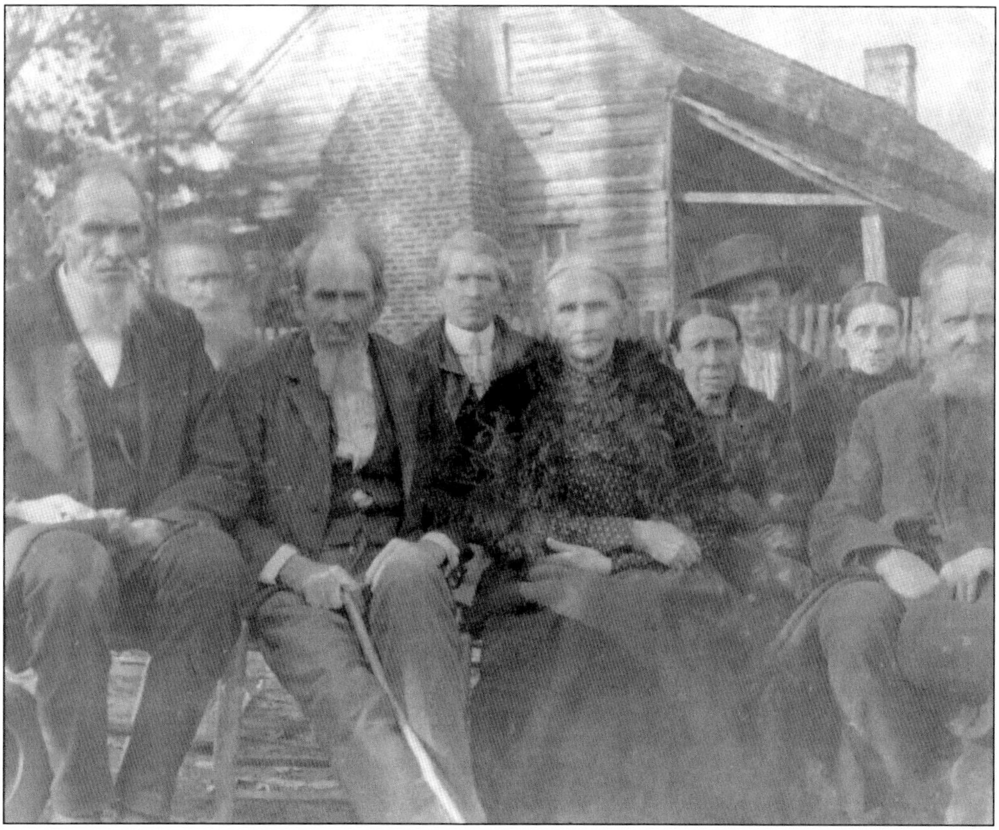

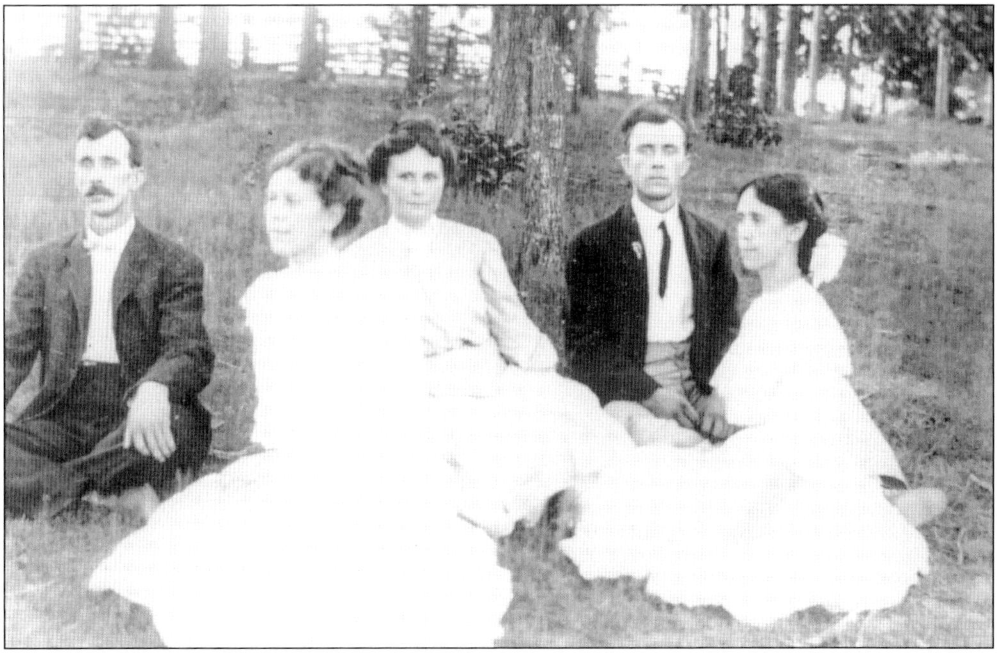

The children of Talmon Harbor (sometimes Harbour or Harber) are pictured here on a hillside on the Harbor family land just east of Tremont. From left to right, they are James B. Harbor, son of Talmon and his first wife, M.A. Lyle; and Talmon's children by his second wife, Martha Ann Evans: Effie Lee Harbor McFadden, Mary Myrtle Harbor Robison, Eddie Aquilea "Rip" Harbor, and Darthua Harbor Hartsell. Talmon Harbor gave each of his children a large tract of land upon which to raise their families.

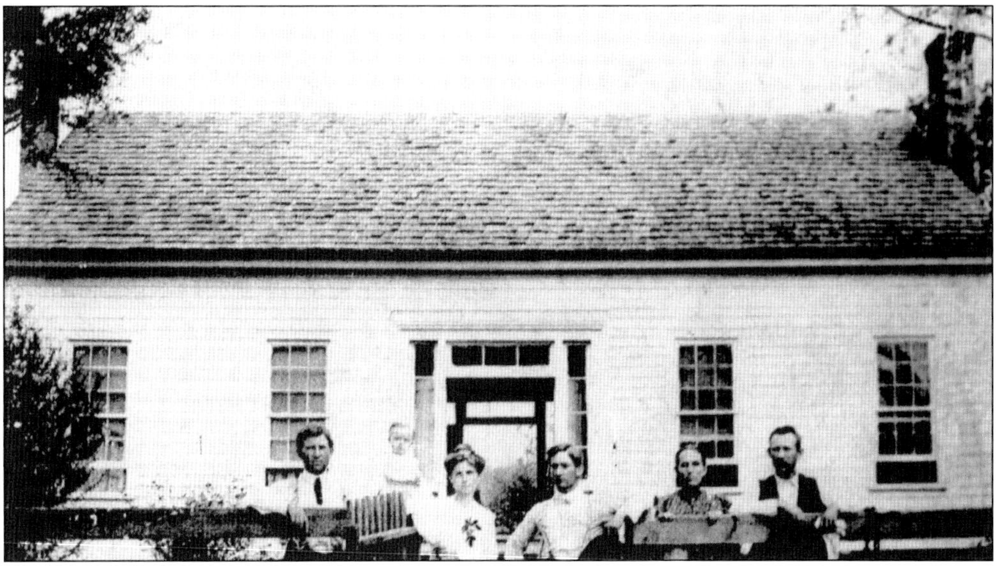

Henry Allison Rutledge returned home from Texas as a widower with two small children and moved into the house begun by his father, Alfred, before he was killed in the Civil War. Standing in front of their home in Van Buren about 1905 are, from left to right, Alfred Ellison Rutledge, son Stacy and wife Louella Boozer, William Greeny Rutledge, Belle Green Rutledge (second wife of Henry), and Henry Rutledge. (Courtesy of Joe Rutledge.)

William Hughes Barkley, left, is shown with one of his brothers in a humorous photograph from the late 1800s. William married Mary Louzetta Hall, and their daughter and only child, Delaphare, was the wife of Sam Mayhall. The Mayhalls, along with their daughter Leatrice Mayhall Timmons, were well-known educators in Itawamba County. (Courtesy of Mary Timmons.)

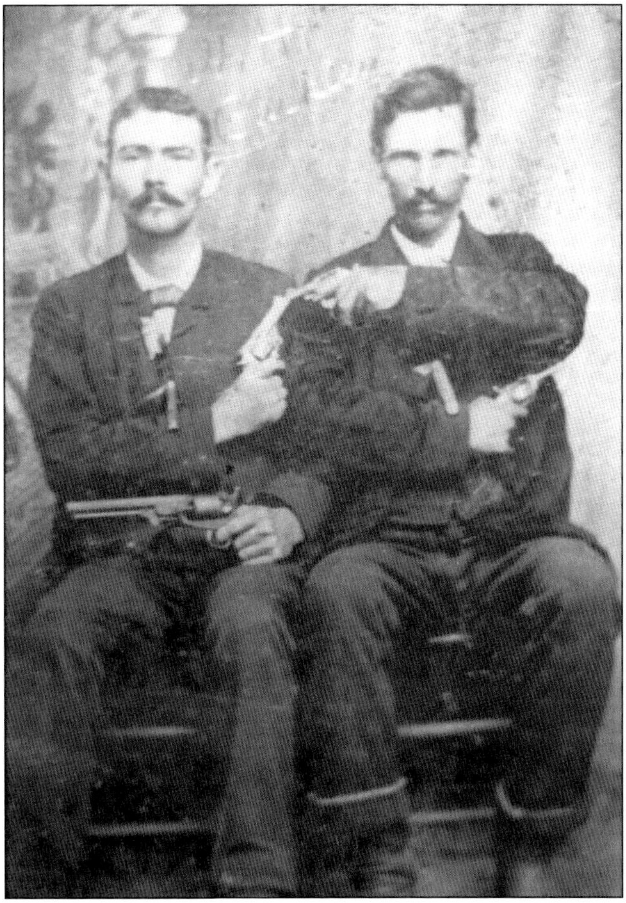

Elected county officials from around 1900 stand against the brick wall of the Itawamba County courthouse. From the county's early days until 1885, the governing board of the county was known as the board of police. The same officers are today known as the board of supervisors.

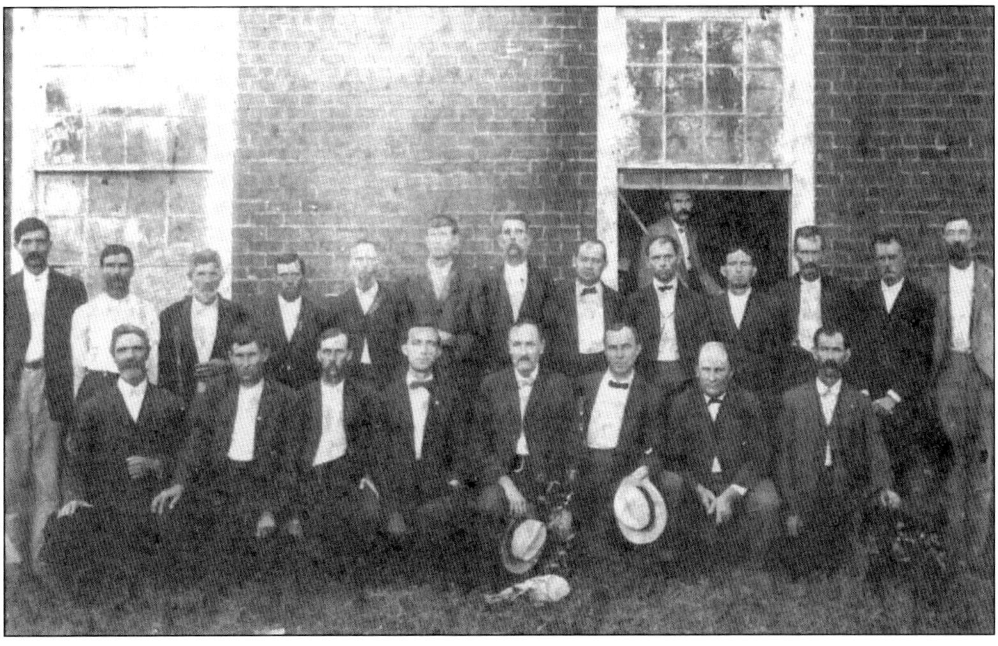

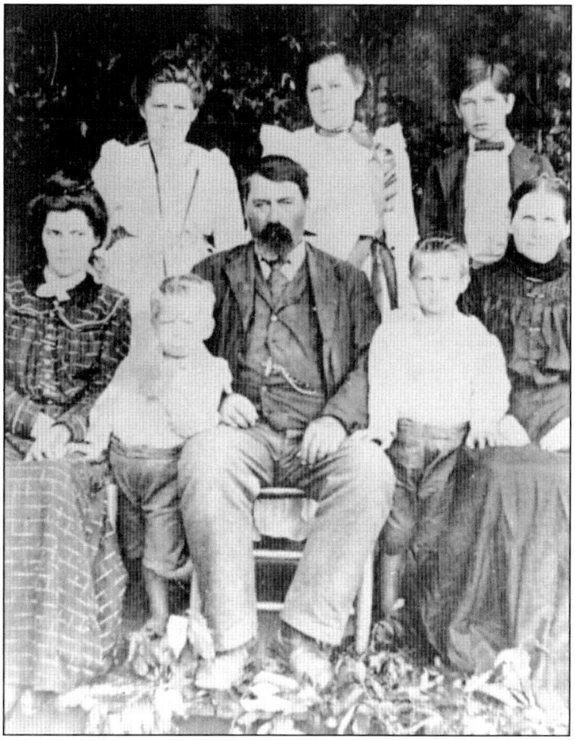

James Solomon Stephens, a farmer by profession, was elected to the Mississippi senate in 1911. In this Stephens family portrait are, from left to right, (first row) Ovie, Dewey, Senator Stephens, Dewitt, and Susan Caroline Shields, the senator's wife; (back row) Essie, Carrie, and Earnest. Not shown are his daughters Sarah, Emma, and Mary by his first wife, Martha Cooper. Senator Stephens was the son of Josiah Stephens and Nancy Midlock Mitchell. (Courtesy of Jonnie Rodgers.)

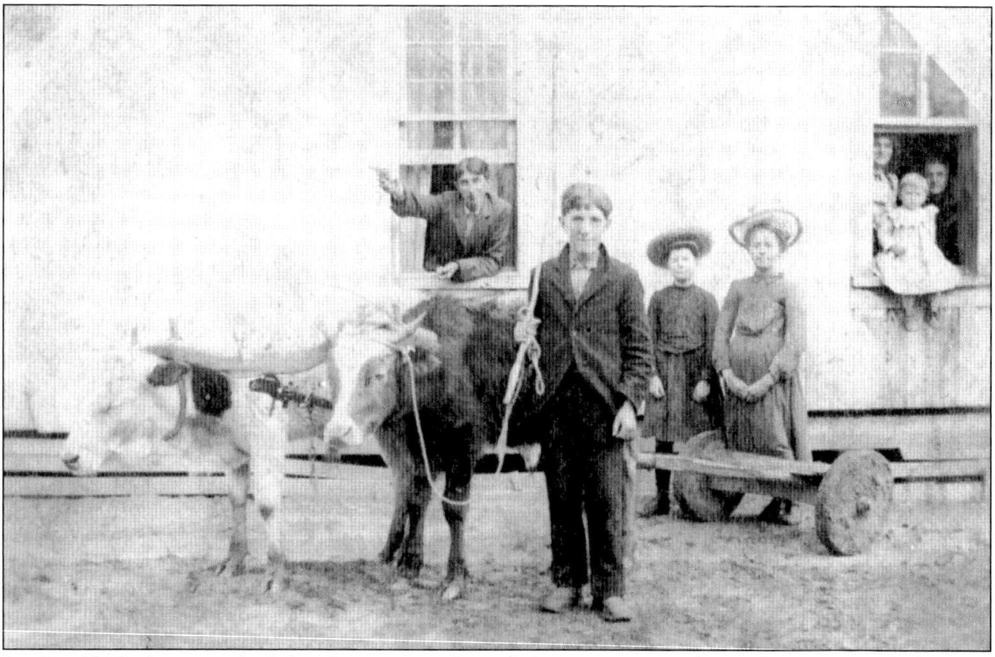

In this 1903 photograph, John Senter stands with his ox team and wooden wagon while Ollie Senter (right) and Audrey Lowery stand by the wagon wheels. In the right window are Celia Senter Griffin (left), Effie Johnson Owens, and Effie's younger brother, Sidney Johnson. In the left window with the pointed gun is Boss Senter. John, Ollie, Celia, and Boss were the youngest four children of Dick and Margaret Waddle Senter.

Cicero Monroe Graham, pictured at right, was in a feud with William "Bill" Dulaney over a saloon Dulaney was operating in Fairview. Graham's daughter, Samantha, was in love with Bill Dulaney's nephew, Henry, but had been forbidden by her father to see him. Samantha's brothers helped her elope in 1883 to Arkansas, where she and Henry settled and began their family. Disowned by her father, Samantha stayed away for 24 years, until, at the request of her dying father, she returned home to the loving arms of her family. The morning after her return, Graham died holding his daughter Samantha's hand. Below, from left to right, are Henry Dulaney, Lillian, Oba D., Samantha Graham Dulaney, and Versey; daughters Myrtle and Minerva are standing in the back. Henry was the son of Thomas Dulaney and Eliza Minerva Horn. (Below, courtesy of Marilyn Leary.)

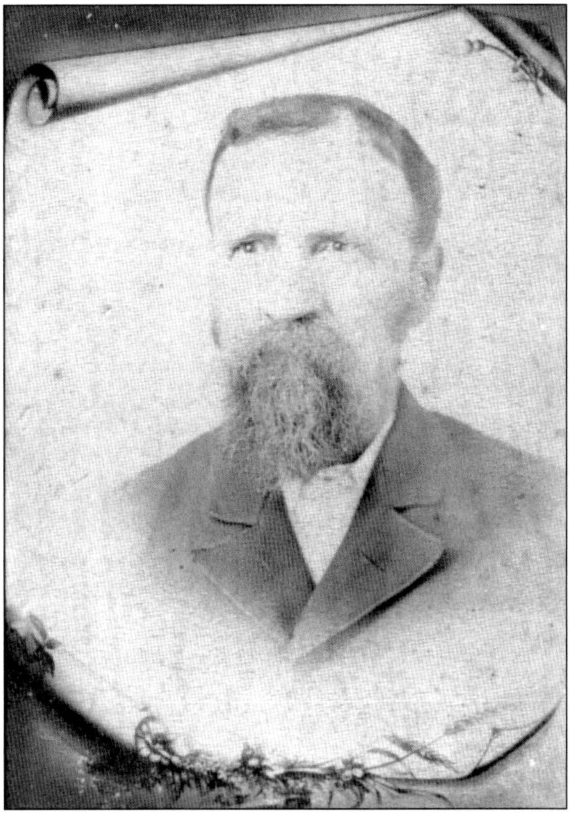

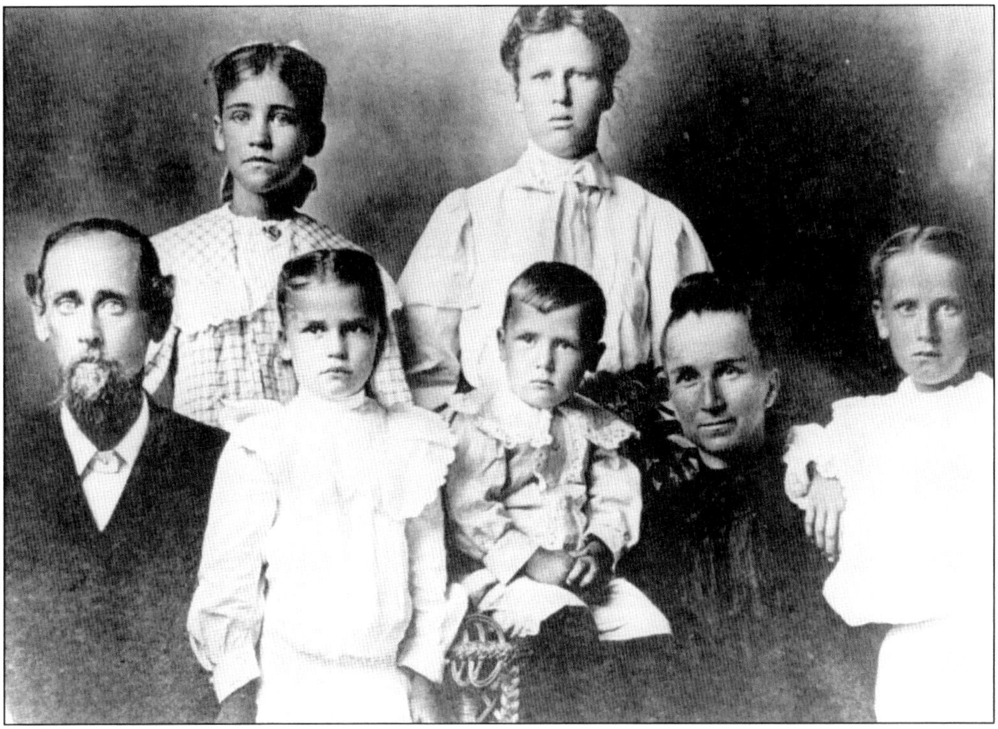

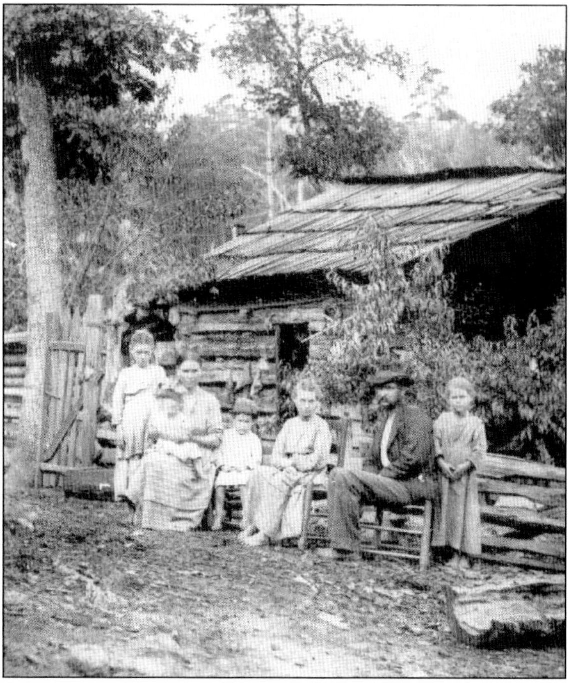

William A. "Bill" and Nila Wiginton Butler lived in the Banner community in northwestern Itawamba County. The 1910 census finds them residing in the Copeland precinct (just above Banner) with their children. Although this photograph was taken after 1900, the log cabin probably dates much older and is a good representation of the earliest living quarters in Itawamba County.

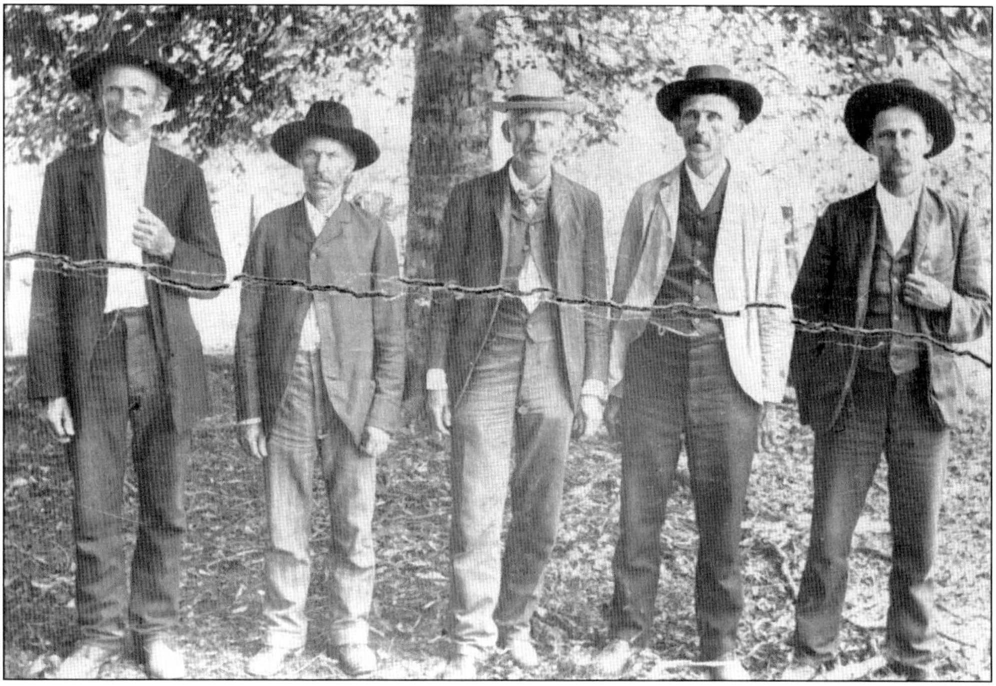

The Clayton brothers were sons of Thompson Clayton and Patience Malinda Brashears, who moved to Itawamba County from South Carolina. Pictured in order of their birth are Nathaniel, Daniel, Elijah, Brooks and Noah. Nathaniel and Daniel lived near Tilden and were known for their woodworking and straight-back, cane-bottomed chairs. Elijah owned the White Springs Resort for several years. Brooks moved to East Tupelo, and Noah was a well-known Baptist minister in Lee and Itawamba Counties.

Getting ready to go hunting with their dogs and guns are brothers Walter (left) and Clifford Rogers. These two young men, along with Trannie Rogers, were the three sons of John Wesley Rogers Jr. and his wife, Mittie Walker. Clifford was married to Celesta Weaver, and Walter died while working in Mexico in 1909.

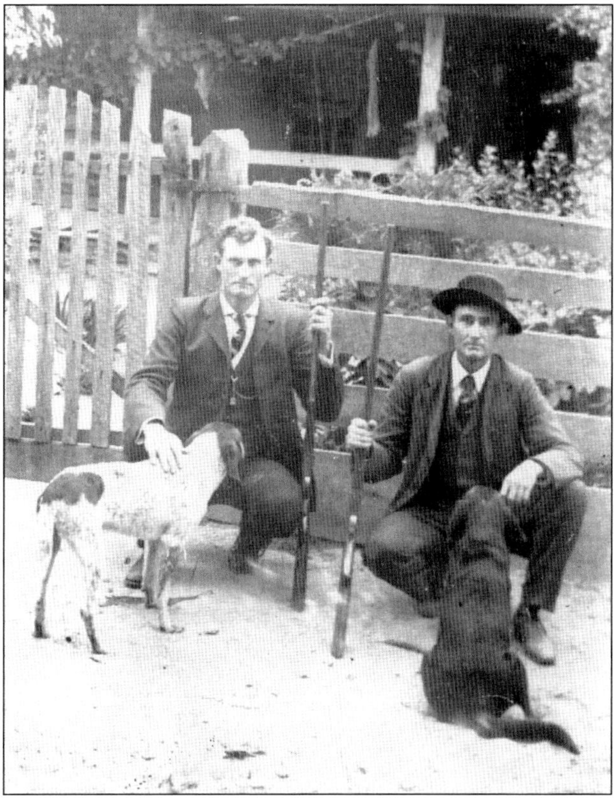

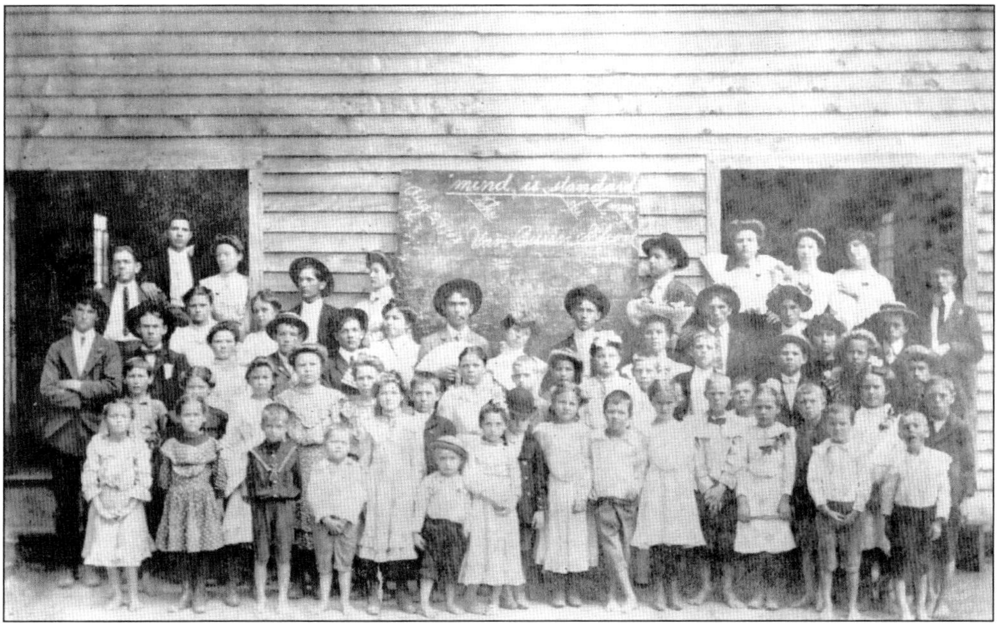

Students at the Van Buren School appear to be putting on a play; one of the male students (center) is smoking a cigar and holding a fan. The date on the chalkboard is August 2, 1904, and diagramed on the board is the sentence, "The mind is the standard of the man," from a poem by Isaac Watts. William Greeny Rutledge was a teacher at the school beginning in 1907.

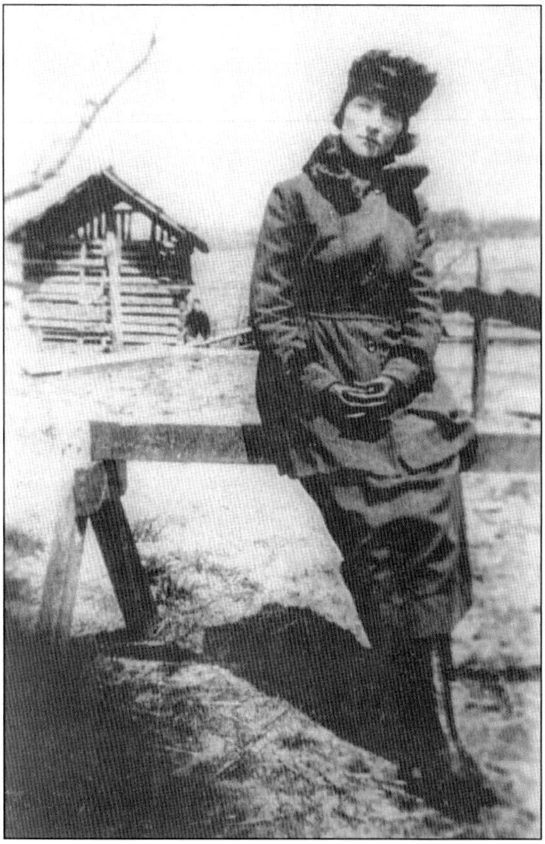

Vista Bernice Graham, looking stylish in 1919 with her fur collar and hat, was born in 1900 in Itawamba County. She was the daughter of Daniel Webster Graham and Emma Tula Farrar of the Fairview community. In 1921, Bernice married Walter Lee Allen, the son of James Henry Allen and Malinda Williams. (Courtesy of Marilyn Allen Leary.)

William and Mary Francis Spencer Toomer made their home in the Clay community, where he farmed, operated a gin and gristmill, and served several terms as supervisor. Standing in front of their home, from left to right, are Reba, Ethel, Emma, Cleta, Lela, Travis, Chester, William, and Mary Toomer. Younger daughters Bulah and Nora are in front of the family. Not pictured is eldest son William Arthur Toomer.

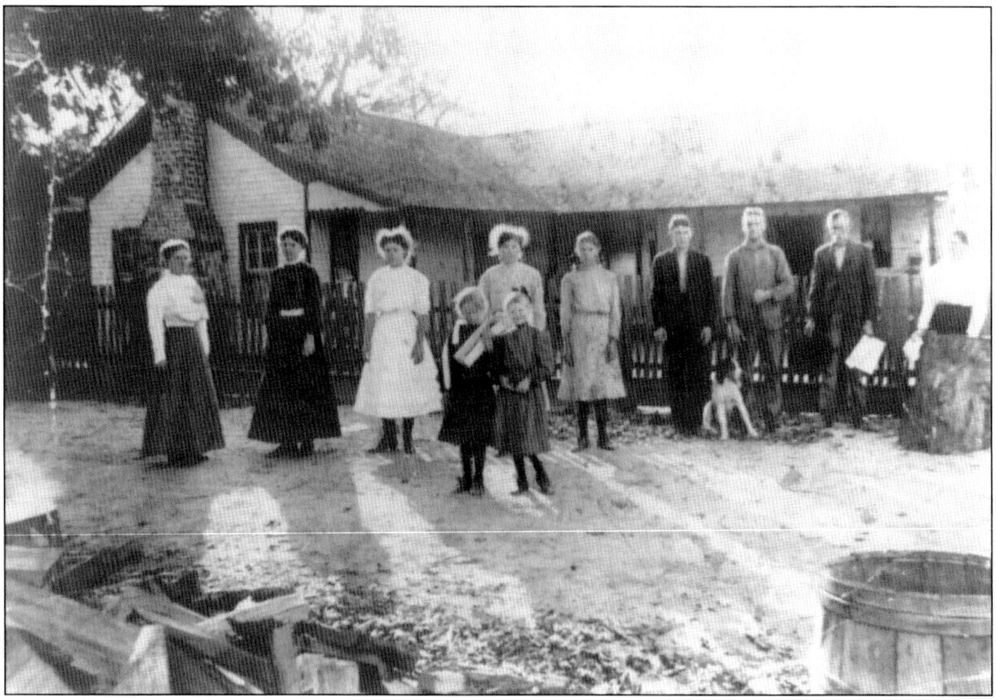

Dr. John T. Senter began his medical practice in 1905 after graduating from Louisville Medical College. He and his wife, the former Carrie Sandlin, are pictured in 1918 with their two eldest children, Lyonel and Graden, both of whom became pharmacists. In addition to the sons shown here, the couple also had a son, Frank, and daughter, Marguerite.

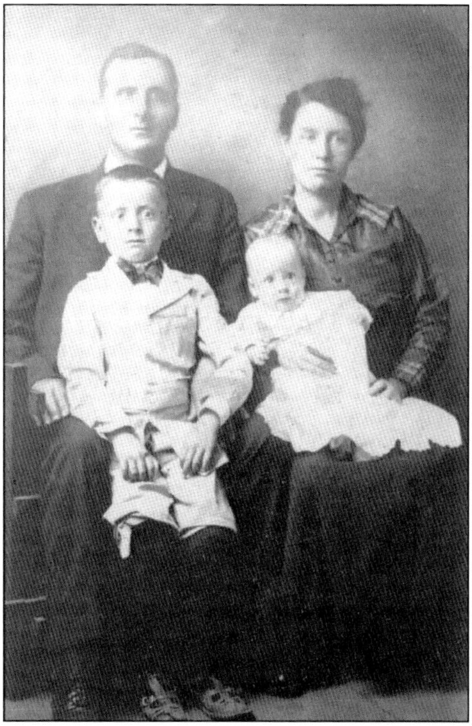

Robert Sidney Sheffield (front and center) is shown with his children and their families in the 1920s at his home in the Dorsey community. Sheffield, the son of Isaac Judson and Fannie McBride Sheffield, was born in 1872. In 1894, he married Leona Gregory, the daughter of Alfred and Sarah Mann Gregory. An educator by profession, he was elected in 1915 to the state legislature. (Courtesy of Joe Nell Wood.)

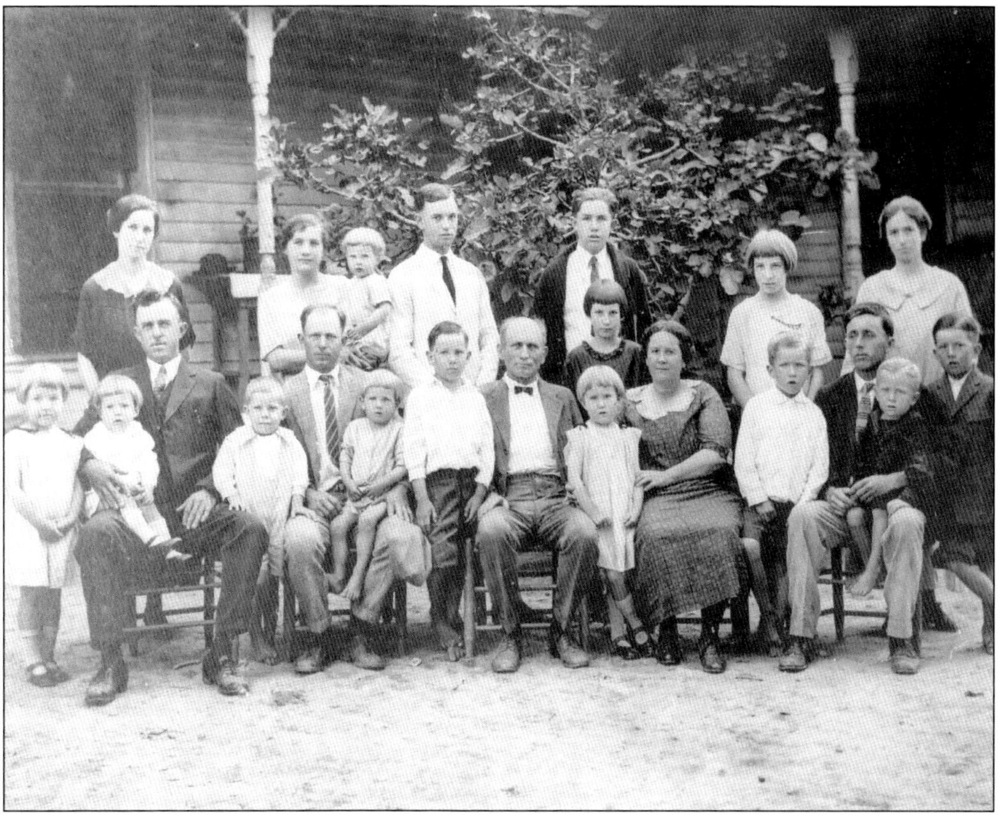

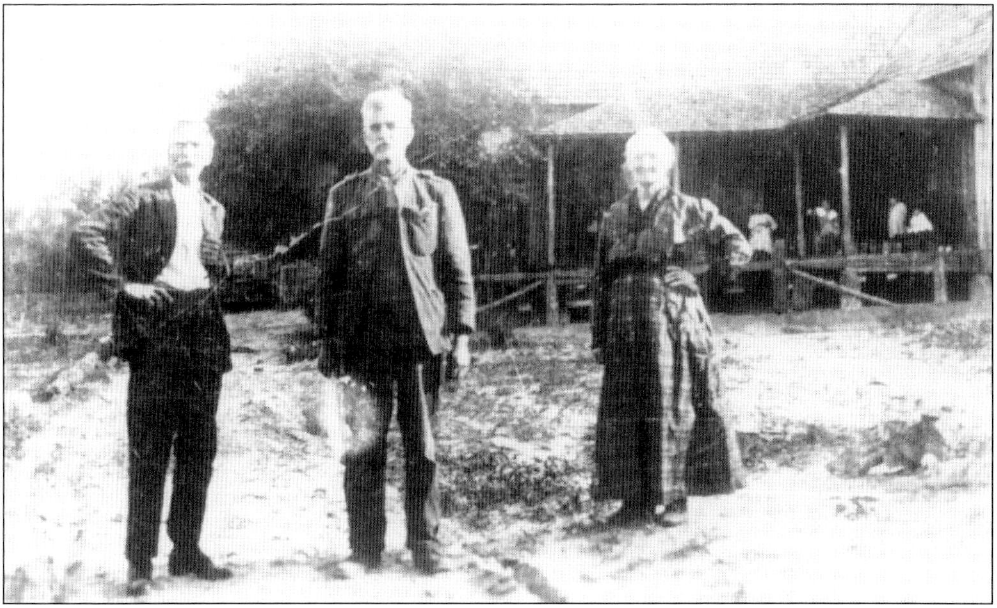

Jackson Samuel Sloan (standing center) moved to the Peaceful Valley community of Itawamba County around 1878 from neighboring Lamar County, Alabama. Jackson married Malissa Caroline Potts (right), daughter of Frances Jane Roland Potts and Jesse Potts, who died in the Civil War. To the couple's left is Malissa's brother, James McFerrin Potts, who also lived in the area.

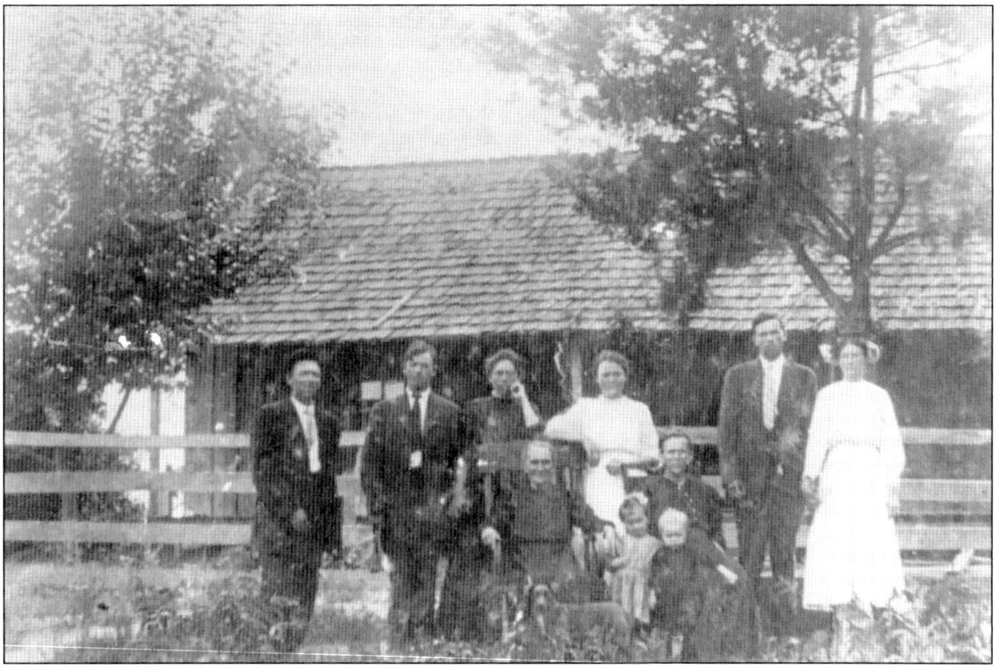

William Greenberry Raburn, who lived at Mantachie, served a term as tax assessor for Itawamba County in the 1880s. Part of his family is pictured here around 1918. They are, from left to right, (seated) Greenberry's widow, Eudoxie Davis Raburn, and her daughter Elvira Raburn Russell with her children; (standing) Frank Raburn, Oscar Younghanse, Ila Raburn Younghanse, Susie Raburn, Ben Raburn, and Nora Berryhill Raburn.

A prominent couple in Tremont was Capt. John Henry Stone and his wife, Florence Cowden Stone. Captain Stone, who received his title from service during the Civil War, was a large landowner. In addition, Stone was a postmaster and a two-term sheriff of Itawamba County and also owned a store, cotton gin, and gristmill. There are many Itawambians who descend from this couple's large family of 11 children.

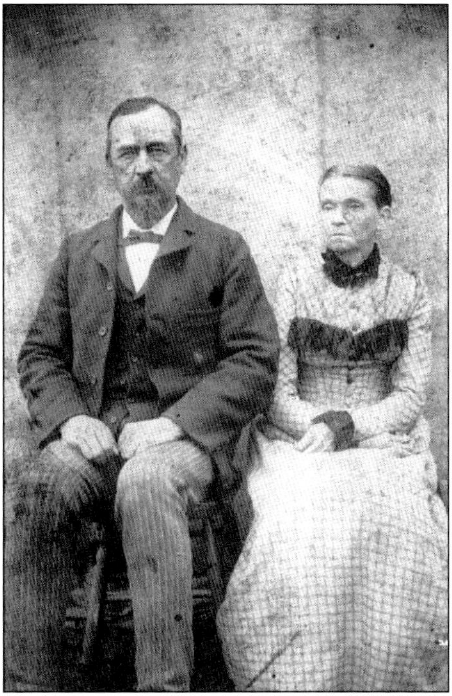

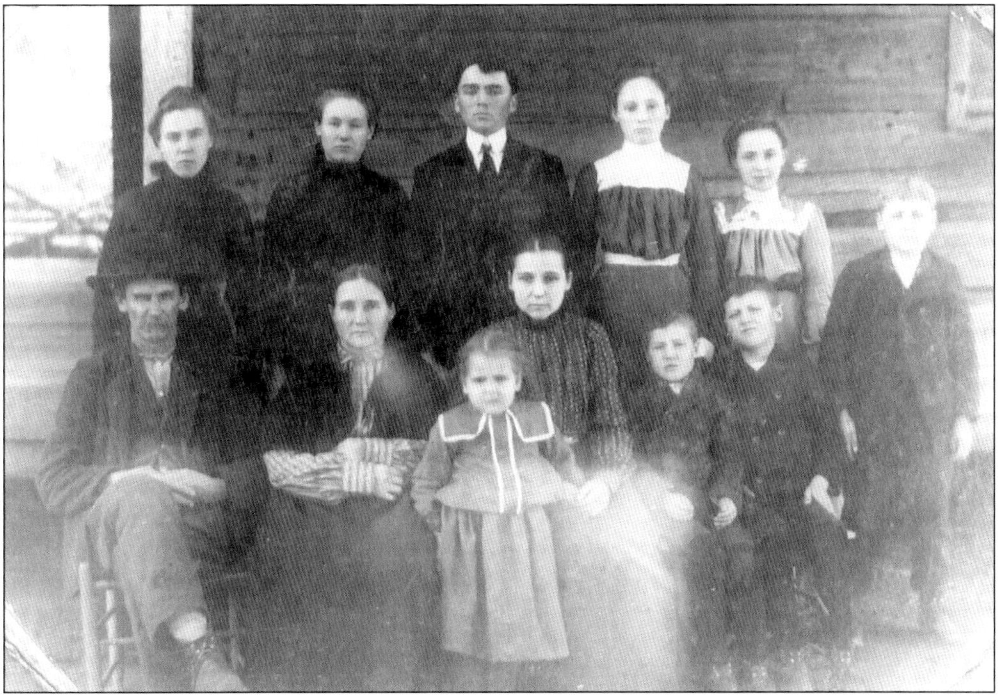

Seated in the front row in this 1902 photograph are, from left to right, Joseph Dulaney with his wife, Martha Ann Johnson, and Rachel Dulaney Williams with her children, Lillian, Alvin, and Ben. Standing in the back row, from left to right, are Ophie Dulaney Jarrell, Willie Jane Dulaney Works, Joe Abb Dulaney, Florence Dulaney Steele, Belle Dulaney Tucker, and Charlie Dulaney. Not pictured is Leona Dulaney Palmer. (Courtesy of Esther Tucker Sheffield.)

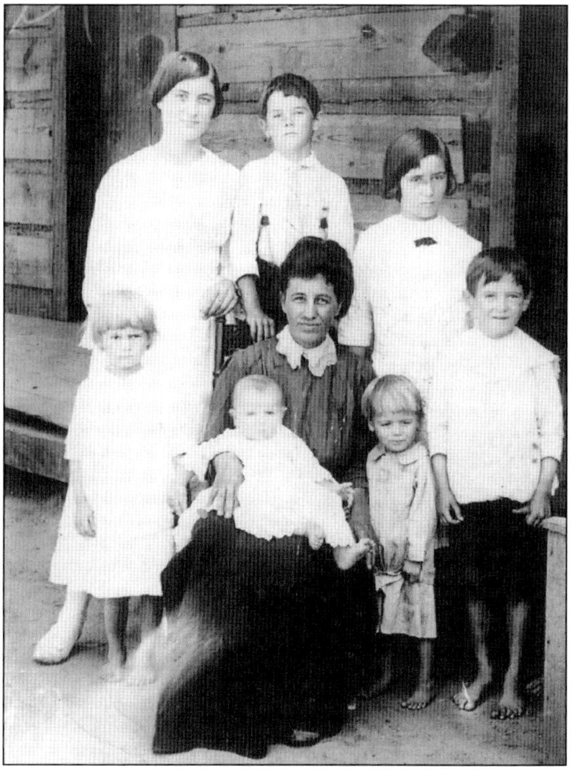

Photographed at her home in Fairview in 1916 is Martha Alice Conwill Taylor with her children. Martha, the daughter of Jesse and Amanda Lentz Conwill, is holding her son Golden Cread Taylor in her lap, while the two small children flanking her are Ester Taylor Thomas (left) and Troy Taylor. Standing behind their mother are, from left to right, Lela Taylor Leathers, Elmer Taylor, Gladys Taylor Dulaney, and Audie Taylor. (Courtesy of Johnny Taylor.)

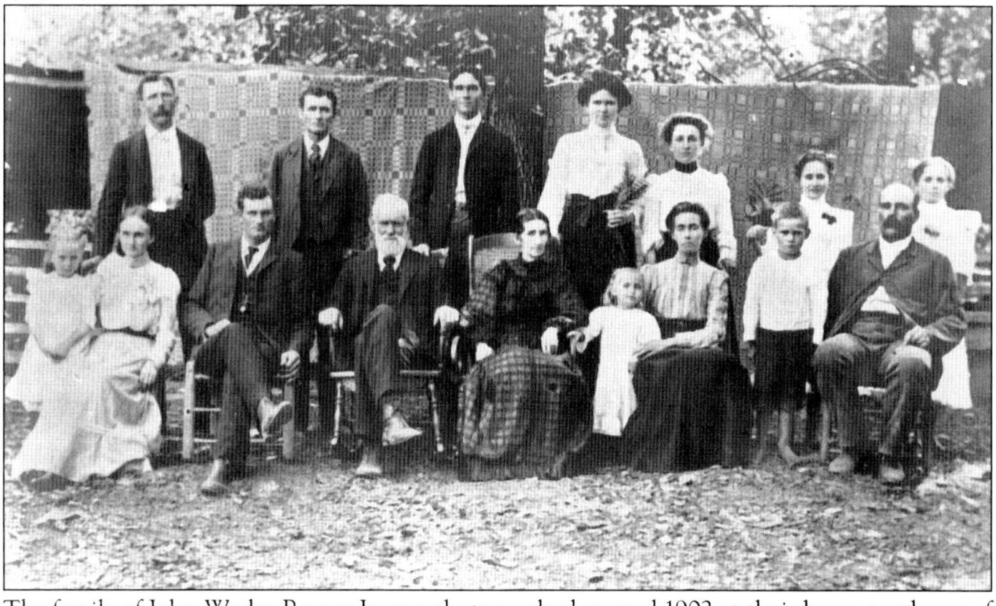

The family of John Wesley Rogers Jr. was photographed around 1902 at their home southeast of Fulton. They are, from left to right, (first row) Charlie Mae Rogers Pierce with daughter Velora, Walter Rogers, John Wesley Rogers Jr., Mittie Walker Rogers, and Sallie Rogers Bowen with her daughter Lula, son Telpha Bowen, and husband James Bowen; (second row) Porter Lamar Pierce, Clifford Rogers, Trannie Rogers, Linnie Lee Rogers, Valera Bowen, Melvina Bowen, and Della Bowen.

Harrison Hood, son of Joshua and Margaret Johnson Hood, married Lydia Minyard, daughter of Thomas and Martha Lesley Minyard, in 1880. The couple is pictured here about 1910 with their children Charlie (seated left of father) and Eula (standing between parents). Standing tall in the rear is Richard, the next tallest is Dennis "Den," and the two small boys are Durell and Ellie.

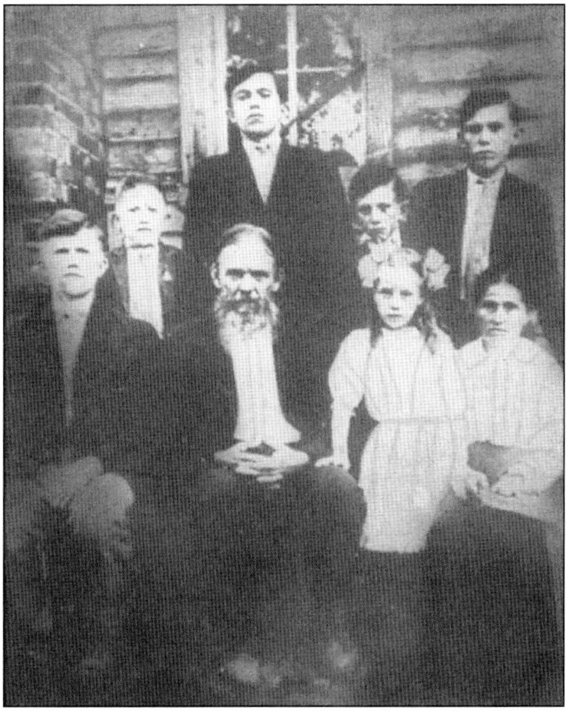

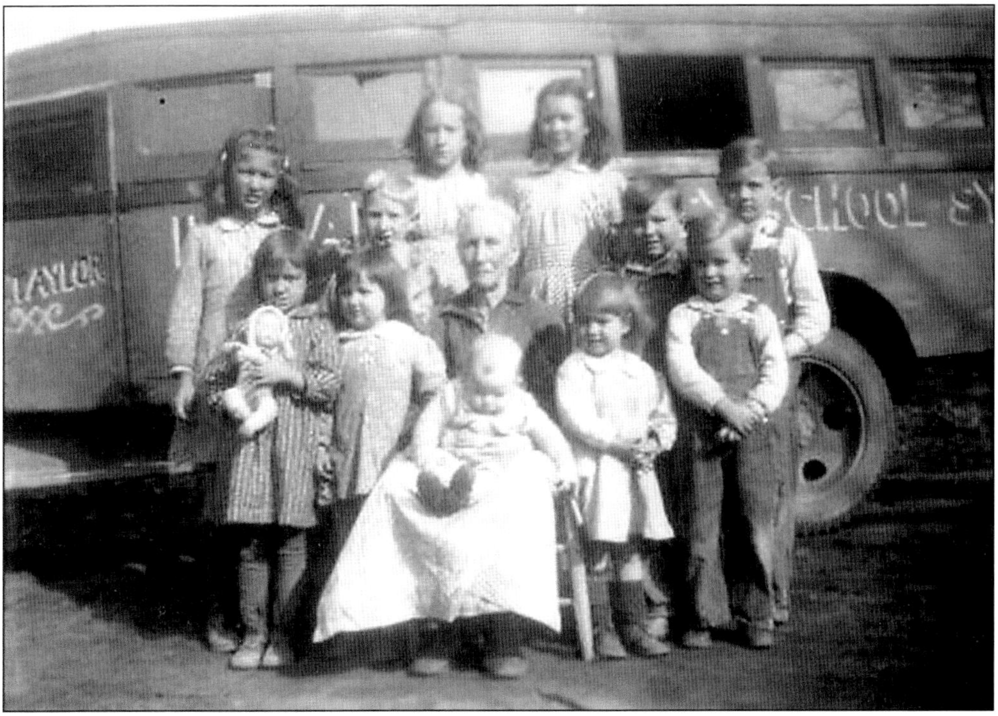

Nancy Elizabeth "Betty" Thornton, daughter of James and Sallie Evans Griffin, was over 80 years old when she sat for this photograph in 1940, surrounded by her great-grandchildren in front of Lawrence Taylor's school bus. John T. and Betty Griffin Thornton's children were James, Fatina, Carrie, Oscar, Mary, Lula, Willie, Nora, Thomas, Lilly, Ira, Belva, and Dewdrop.

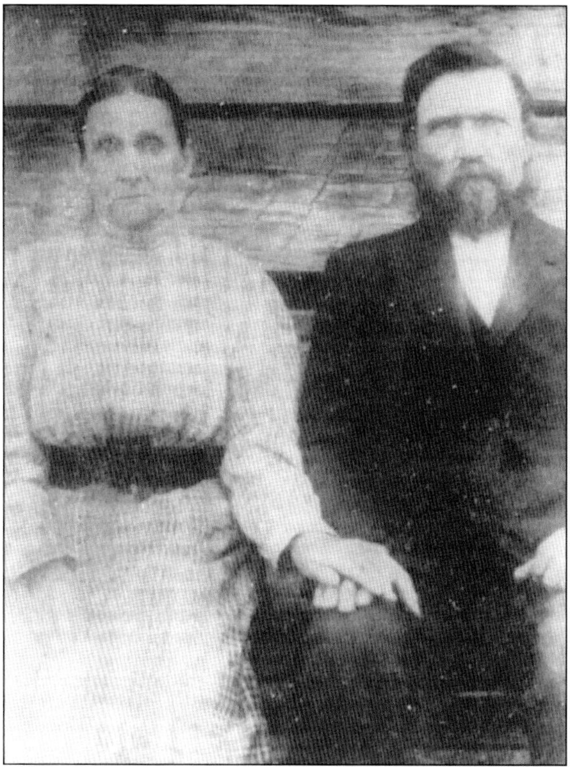

George and Mary Dulaney West lived on a farm in the Dulaney Branch community. George was the son of Simon and Nancy Thompson West, while Mary Elizabeth was the daughter of Alfred and Rachel McNiece Dulaney. The West and Dulaney families were among the earliest settlers of the county. (Courtesy of Virble West Booth.)

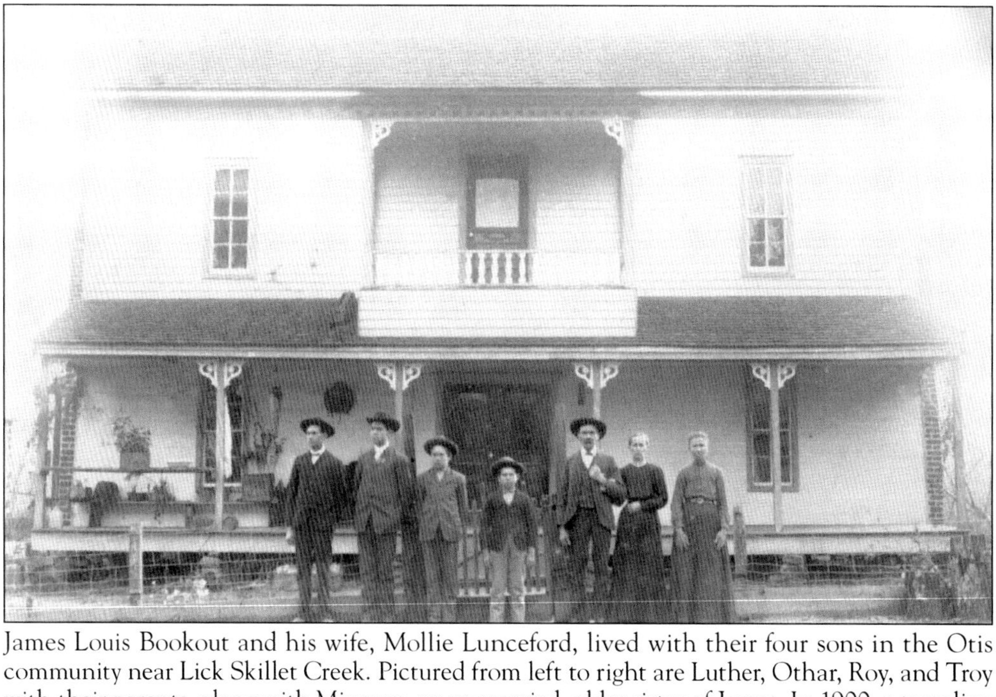

James Louis Bookout and his wife, Mollie Lunceford, lived with their four sons in the Otis community near Lick Skillet Creek. Pictured from left to right are Luther, Othar, Roy, and Troy with their parents, along with Minerva, an unmarried, older sister of James. In 1900, a traveling German came through the area, and, living with the Bookout family for several months, painted the home with his hand-mixed paints. The paint job lasted nearly 50 years.

Cora Desta Reich poses on one of the county's iron bridges. The daughter of John Coatland and Emily Weaver Reich, Desta met and married Seth Harvey Alexander, a Texas native, when he was working on the Bankhead Highway in the 1930s. They had one daughter, Janie. The Reichs were of German ancestry. (Courtesy of Janie Alexander.)

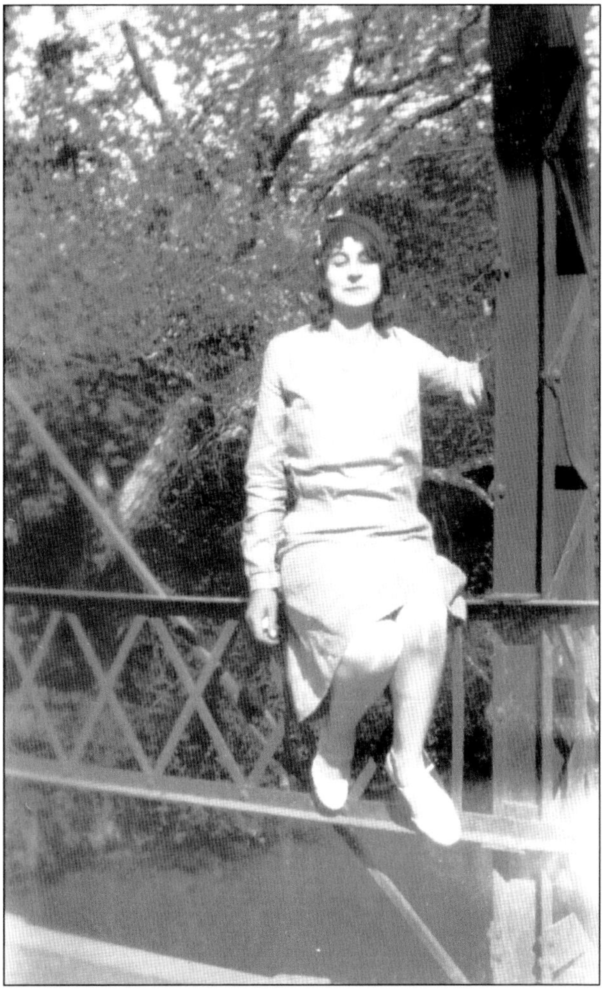

Charles Andrew Dozier sits in his 1919 Model T car, one of the first cars on the streets of Fulton, with his wife, Lucinda. Surrounding them are their children, from left to right, Oren, Orville, Charleen, and Beulah. Children not pictured were Tula and Mae Ola. Andy was a well-known freight hauler, carrying freight between Fulton and Tupelo for nearly 50 years and using horses and mules in the early days.

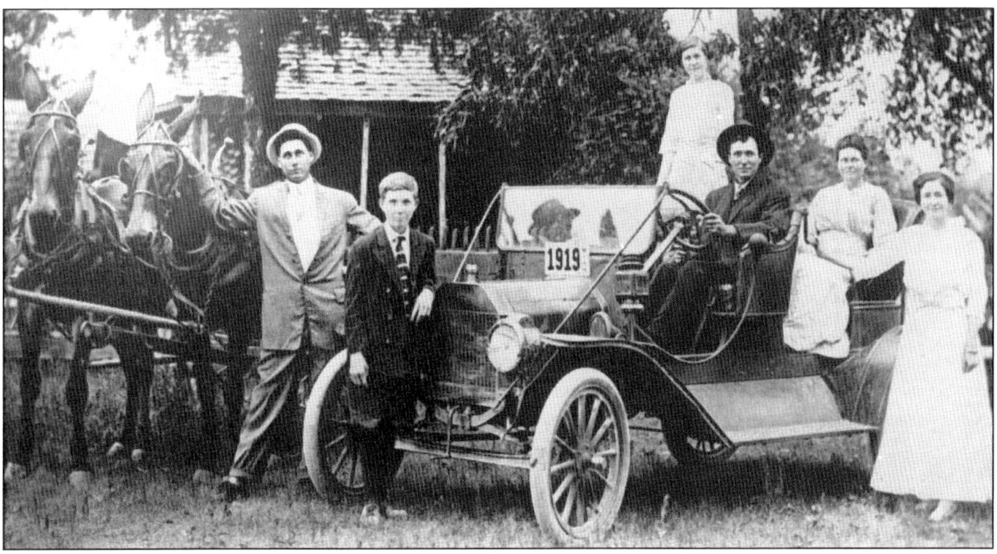

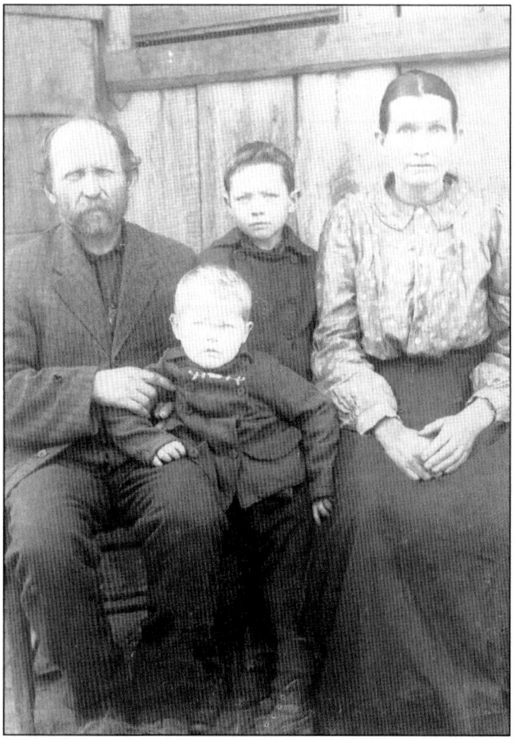

Reuben Wiygul Jamerson was the son of Thomas Jefferson Jamerson, a Delaware native said to have been an herb doctor, and Susan Elizabeth Ross. The Jamersons settled in the Sandy Springs area. Reuben and his wife, Sarah McKee, daughter of James and Emiline Ashley McKee, are pictured here with their youngest two sons, Luther (rear) and Girland. Other children were Grace, Ovie, Lillie Belle, Austin, Lona, Minnie, and Edgar. (Courtesy of Jada Jamerson Mills.)

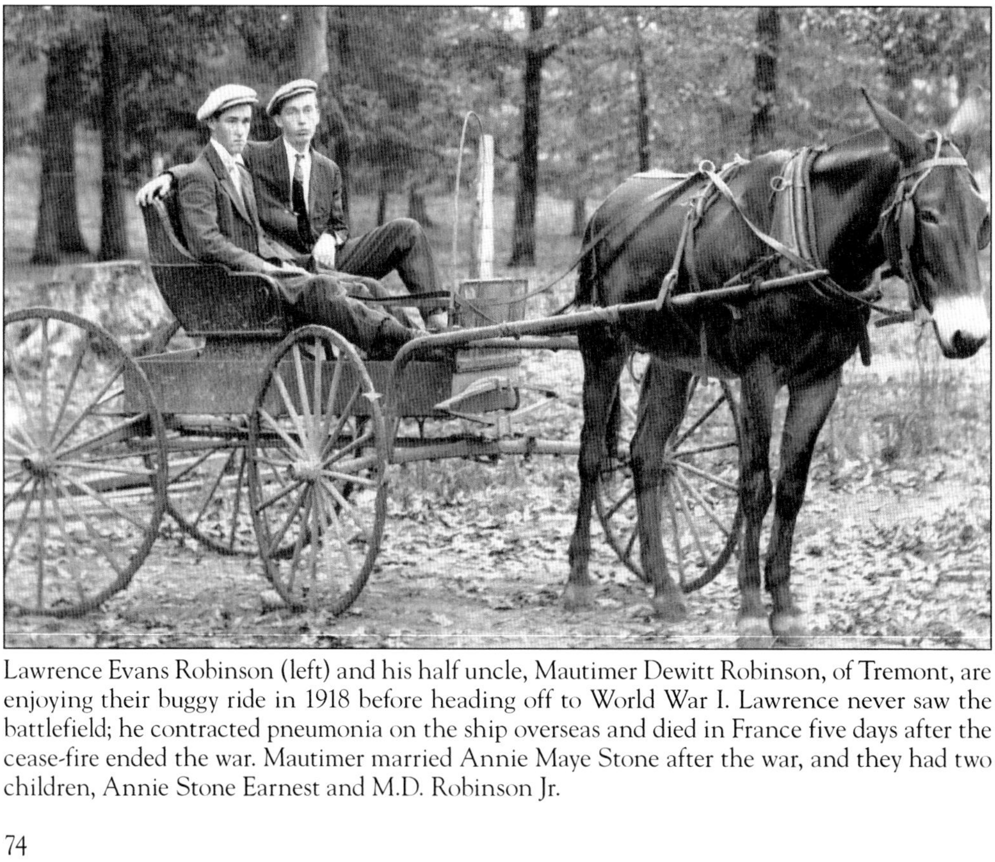

Lawrence Evans Robinson (left) and his half uncle, Mautimer Dewitt Robinson, of Tremont, are enjoying their buggy ride in 1918 before heading off to World War I. Lawrence never saw the battlefield; he contracted pneumonia on the ship overseas and died in France five days after the cease-fire ended the war. Mautimer married Annie Maye Stone after the war, and they had two children, Annie Stone Earnest and M.D. Robinson Jr.

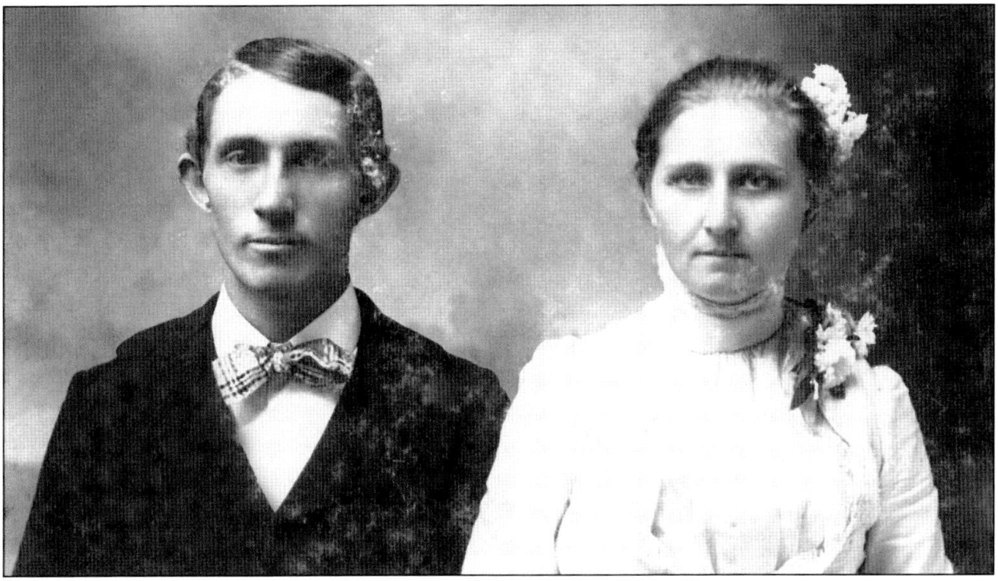

Walter Clifton Cummings married his sweetheart, Tamsie Ophelia Mitchell, on December 3, 1899. Because they were second cousins on the Ratliff side, family and friends were against the marriage. Walter went to Texas to forget about Tamsie, but he could not get her out of his mind, came back to Itawamba County, and married her. This may be their wedding portrait. (Courtesy of Coleen Thornton, great-granddaughter of W.C. and Tamsie Cummings.)

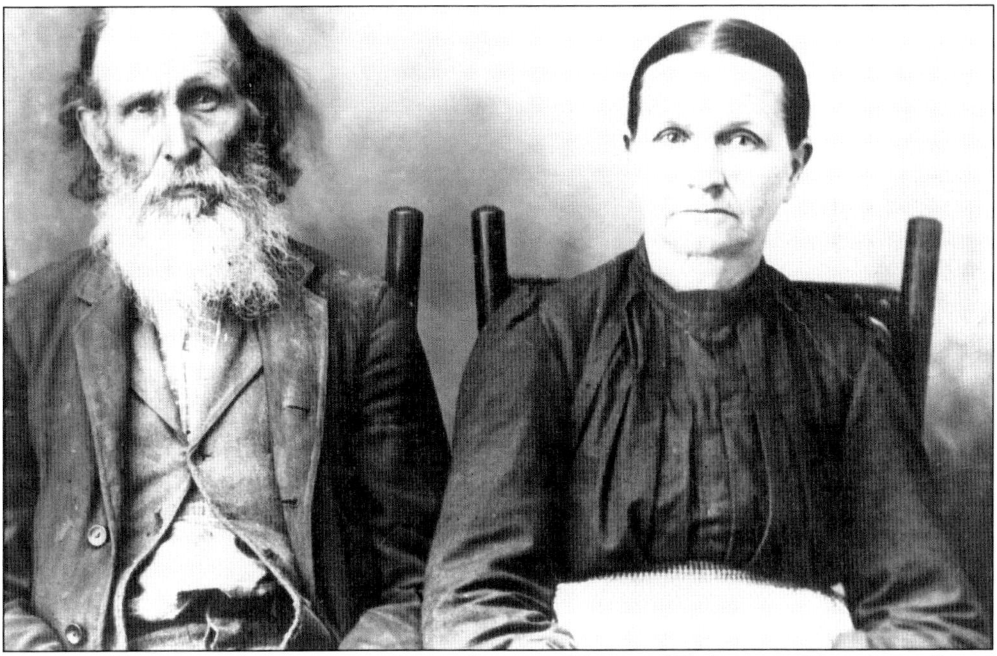

Isaac Brinkley Conwill, born in 1845, was the son of George Washington King Conwill and Elizabeth Prestridge. He married Sarah Elizabeth Parker, the daughter of Micajah Parker Jr. and Martha Matilda Hankins, on November 13, 1867. Isaac and Sarah lived in the Bounds Crossroads area of Itawamba County and are buried at Providence Cemetery.

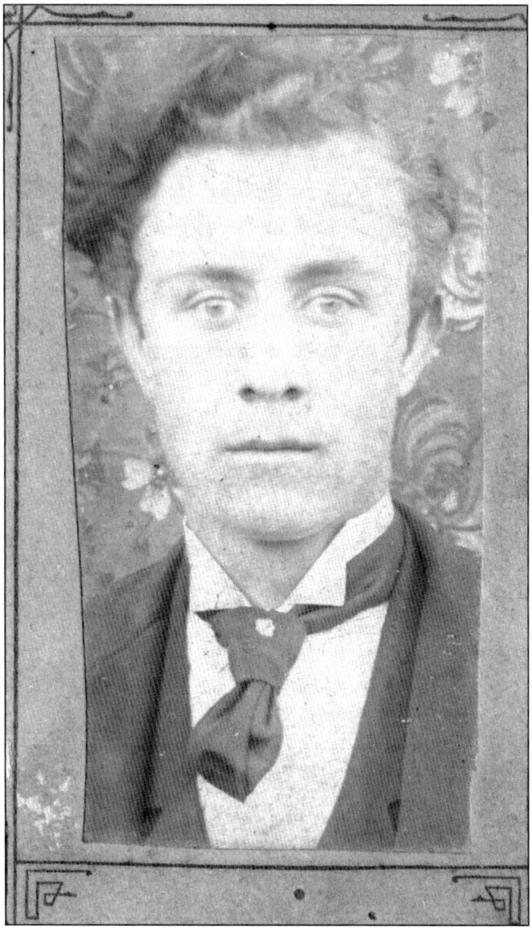

E.A. "Rip" Harbor (sometimes Harbour) was a photographer who lived in Tremont. Several images from around the turn of the 20th century have his name imprinted on the card stock as "E.A. Harber, Photographer." The son of Talmon and Martha Evans Harbor, Rip married Vannah Cofield, daughter of Sam and Zilpha Cockrell Cofield. They had one son, A. Harbor, and raised three of Vannah's orphaned nieces—Ruby, Pearl, and Pauline Cofield.

In this 1890s photograph, the Joseph S. Davis family gathers in front of their home near Fawn Grove. Pictured are, from left to right, (first row) Leana Davis Sheffield holding daughter Anna, Icy Comer, Joseph S. Davis, Archie Comer, Zelma Comer, Martha Brown Davis, Santford Comer (in front), Arthur Comer, and Gala Ella Davis Comer; (second row) William Tom Davis, Isaac R. Davis, Rafe Davis, and Houston N. Davis.

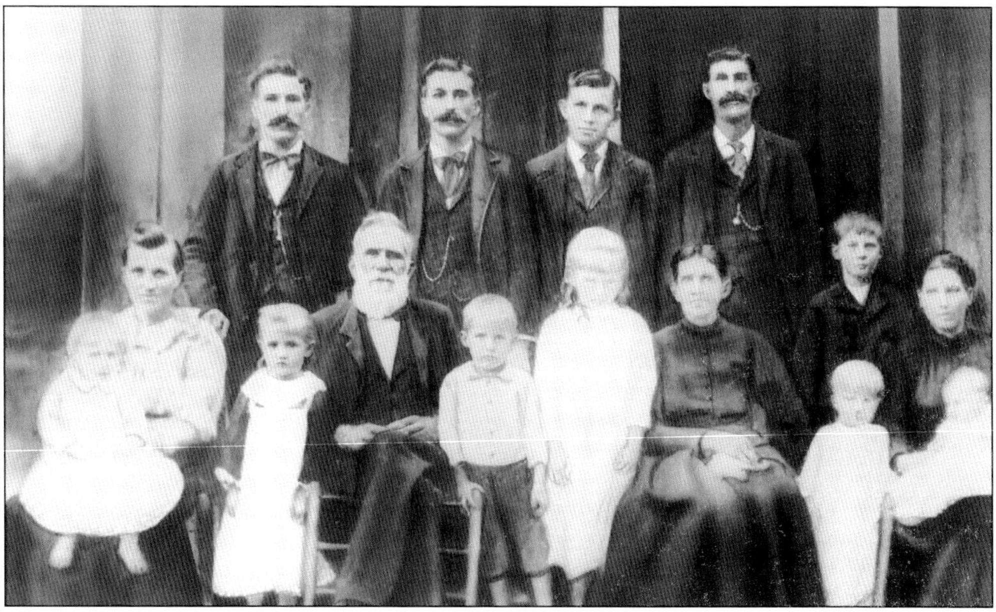

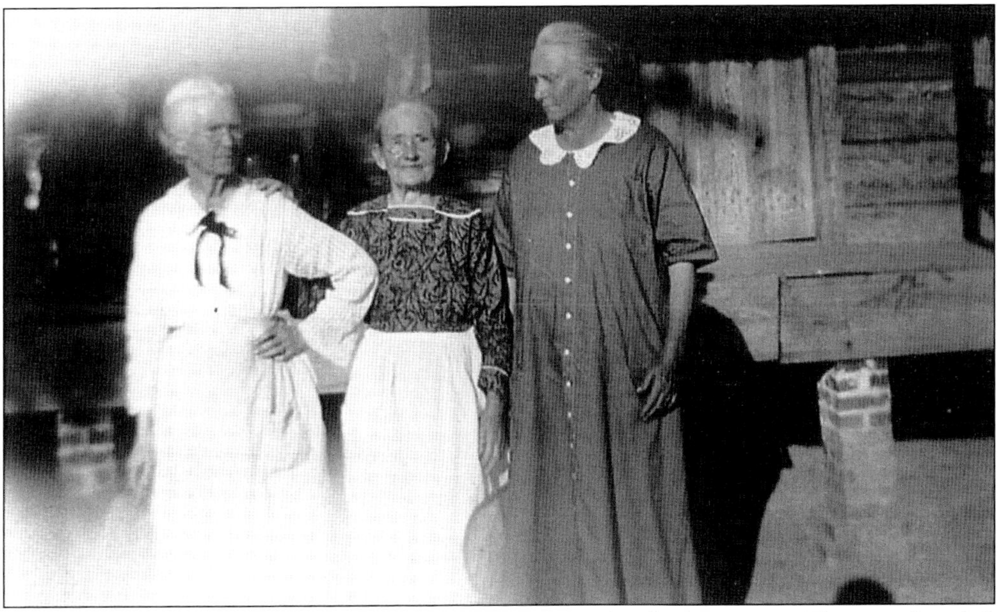

Three ladies of the Cardsville community stop to enjoy each other's company. They are, from left to right, Nancy Neal Cauthern, Edora Elzira Rutledge Bourland, and Rebecca "Becky" Bourland Patterson. All three women lived long lives and are buried near each other in Bourland's Cemetery. (Courtesy of Garry Morris McFerrin.)

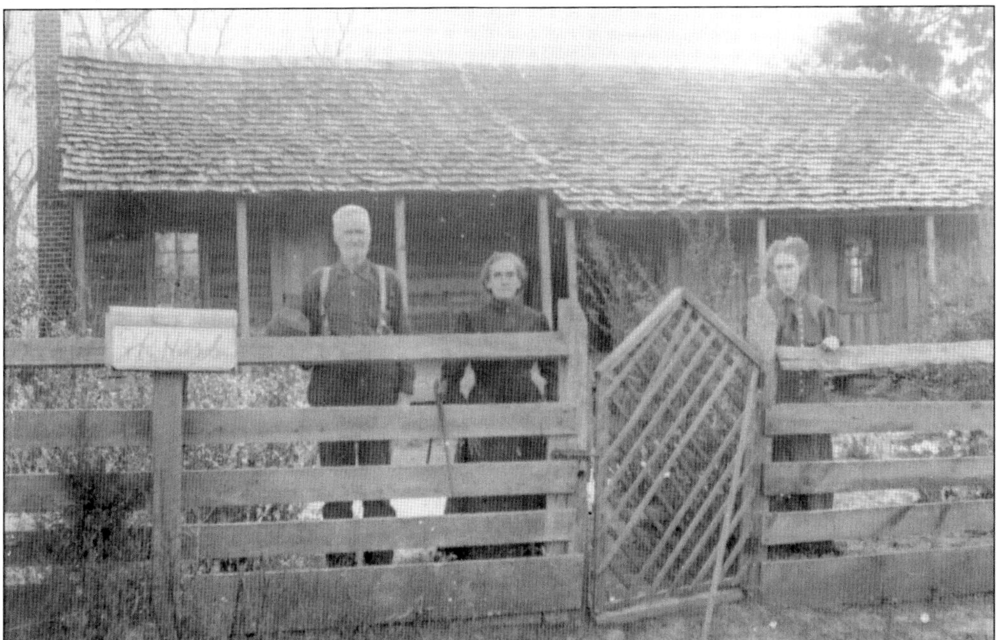

Standing in front of the Harden home are Richard Ivy Harden and his wife, Nancy Lilly Blaylock, and daughter, Ivy Lee Harden. Richard was a blacksmith and farmer, and he owned a thresher and cotton gin. After the Civil War, Richard bought land in the area now known as Harden's Chapel. He and Nancy gave land for a church, originally called Pleasant Grove but later changed to Harden's Chapel Church by the Methodist conference to honor the Hardens.

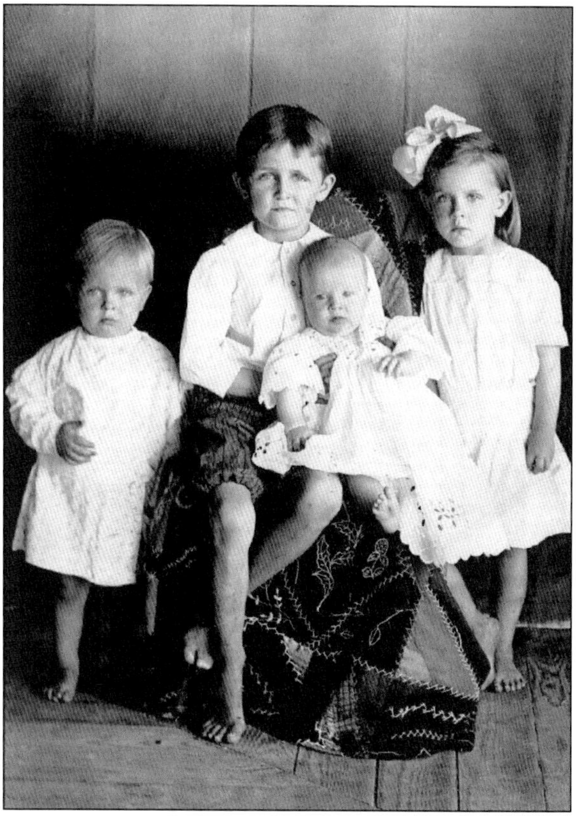

These young Pennington siblings, children of William Hugh and Ethel Dee Sloan Pennington of the Peaceful Valley community, had their photograph taken in 1914. From left to right, they are Fessie Manuel Pennington, eldest son Jesse Gordon Pennington holding little brother Gaylord, and Vivian Irene Pennington. Not yet born were Frelon Jury, Mavis Maxine, and Clara Nell Pennington.

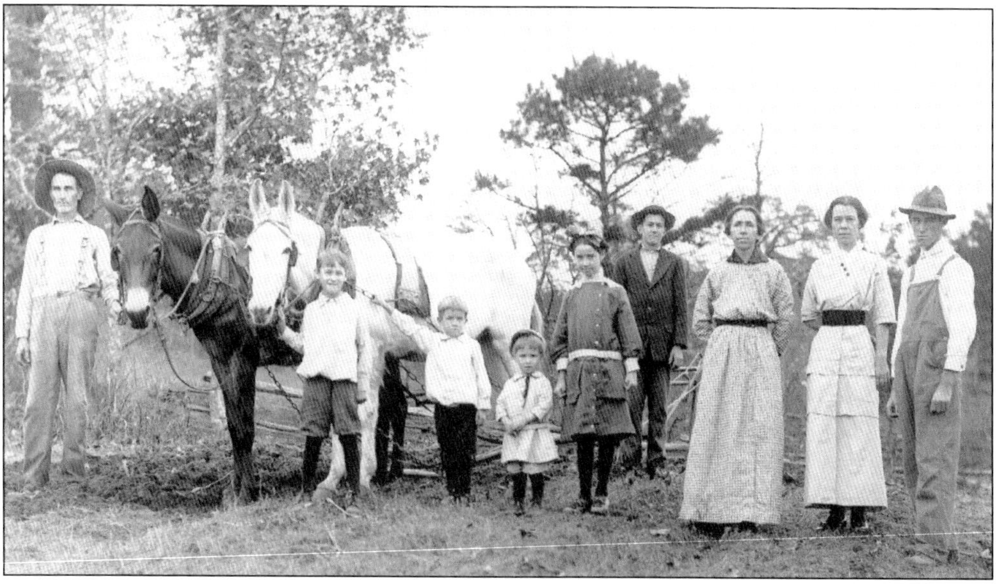

On the family farm around 1914 in Tremont, with their prized mules are, from left to right, Gideon Casibay Robinson with children Luke Lee, Louis Gideon, Lowrey Marlin, Lilly Jewell, William Lawson, wife Arthusa Parneshia "Thusie," Lilly Evans (Thusie's sister), and eldest son Lawrence. Around 1928, Gid and Thusie moved to Fulton, where they farmed on their land that was located where Willow, Mimosa, and Farrar Streets are now.

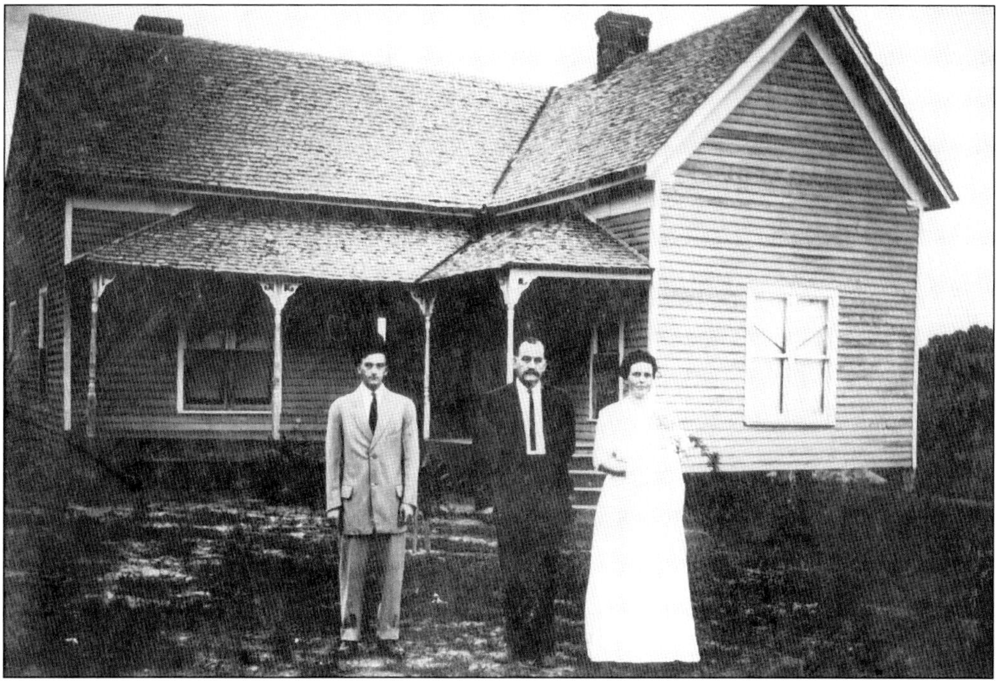

Standing in front of her home in 1913, Josie Sims is holding her eldest son, Loyce, in her arms. Josie's husband, Joseph Sims (center) and stepson, Yancy, stand beside her. Josie, daughter of Steve and Emily Bigham Turner, taught 34 years in the Mantachie schools. She married Joseph E. Sims in 1912, and they had two sons, Loyce and Steve, and a daughter, Susie Mae Sims Earnest, who taught English at Itawamba Junior College for many years.

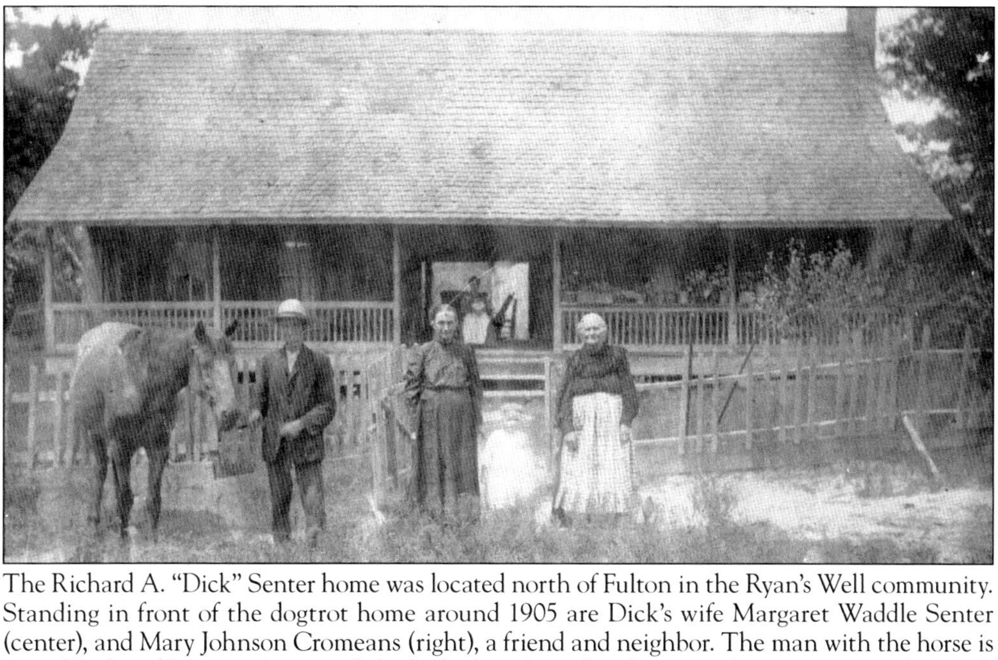

The Richard A. "Dick" Senter home was located north of Fulton in the Ryan's Well community. Standing in front of the dogtrot home around 1905 are Dick's wife Margaret Waddle Senter (center), and Mary Johnson Cromeans (right), a friend and neighbor. The man with the horse is one of Dick and Margaret's sons, while the little girl standing between the women is unidentified, as is the woman on the porch.

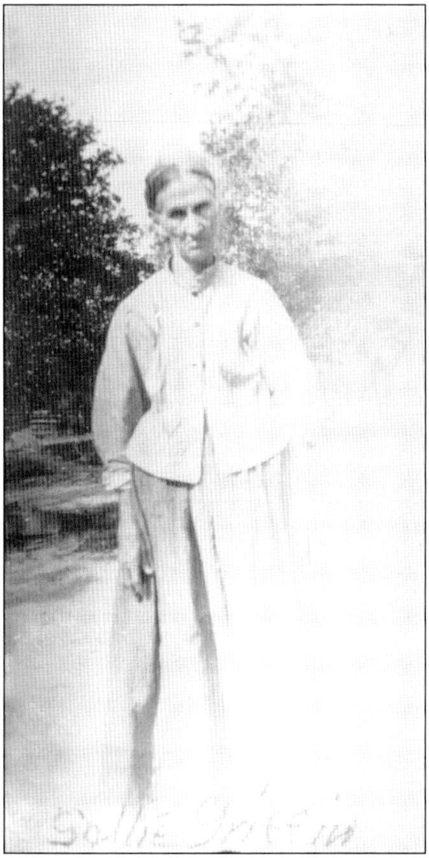

Sallie Evans Griffin was nearly 90 years old at her death in 1924. She was the daughter of John and Meeky Crawford Evans and the wife of James Griffin, who died in 1862. The Evans family owned land surrounding the busy intersection of the Fulton-Iuka and Baldwyn-Red Bay roads. The settlement was known as Evans' Well due to a freshwater well at the intersection. When Ryan's Store opened next to the well and established a post office in 1852, the post office was named Ryan's Well.

The five eldest children of James Henry and Bertha Warren Dulaney hold hands for this photograph in 1923. Standing, from left to right, are Lawrence Orr, Vada Marie, Vida Chesteen, James Melvin, and James Elgar. Not yet born when this picture was taken were Lester, Cotton, and Clarence. The Dulaney family lived east of Fulton, where Jim Dulaney was a country veterinarian and mule trader.

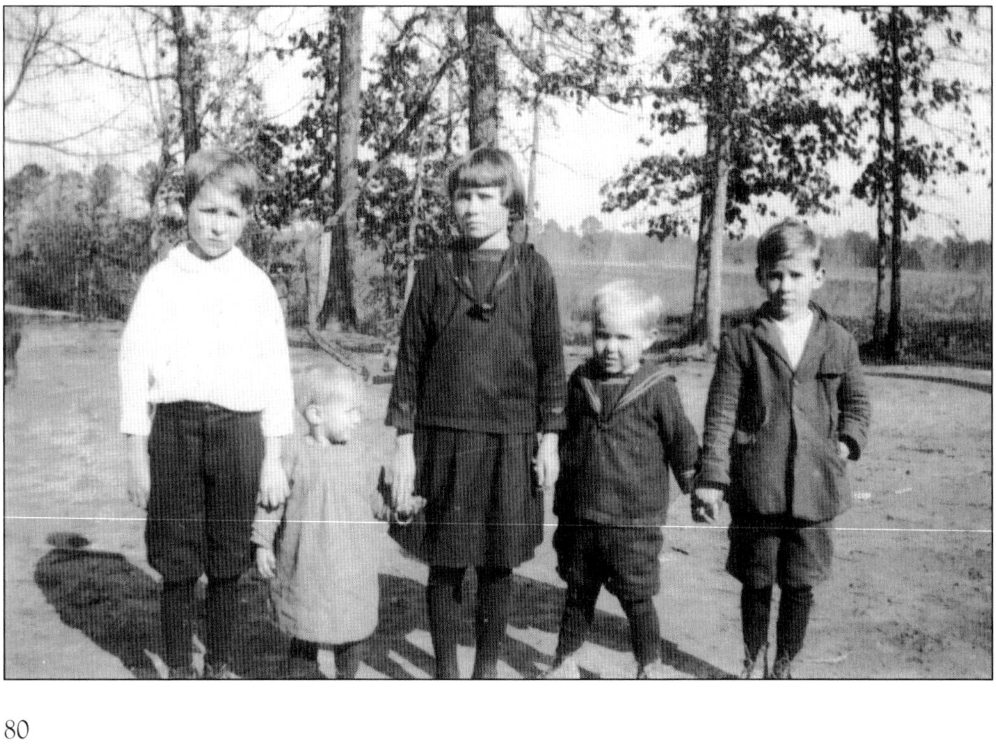

Son of Marcus de Lafayette Spearman and Sarah Green, Paul D.P. Spearman was a Tremont native who became an important figure in the newly emerging communications industry in Washington, DC. In 1929, Spearman coauthored *State and Municipal Regulation of Radio Communication*, and in 1934 he was appointed the first general counsel for the Federal Communications Commission. Spearman was the founder of the firm Spearman and Roberson and served as attorney for the National Association of Regional Broadcast Stations. Below, children from the Stone and Spearman families of Tremont line up for a picture together. They are, from left to right, (first row) Sula Spearman, Olean Spearman, Grace Stone, and Ollie Spearman (second row) Fred Stone, Paul D.P. Spearman, Annie Maye Stone, Crew Spearman, and Owen Spearman. Fred, Annie Maye, and Grace were children of William and Pearl Evans Stone, while the Spearman siblings were children of Marcus de Lafayette Spearman and his two wives, the second wife being Mary Lillian Wiginton.

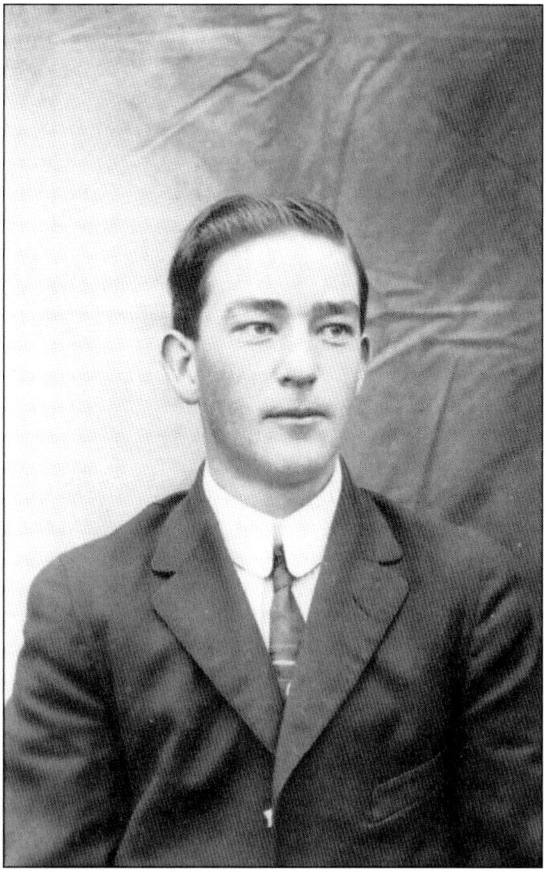

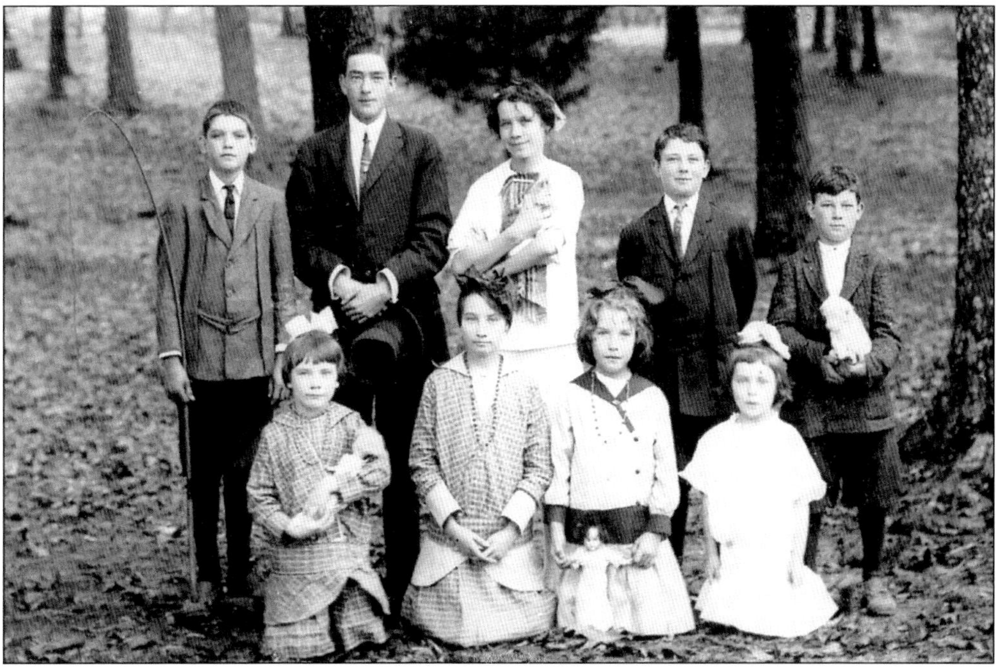

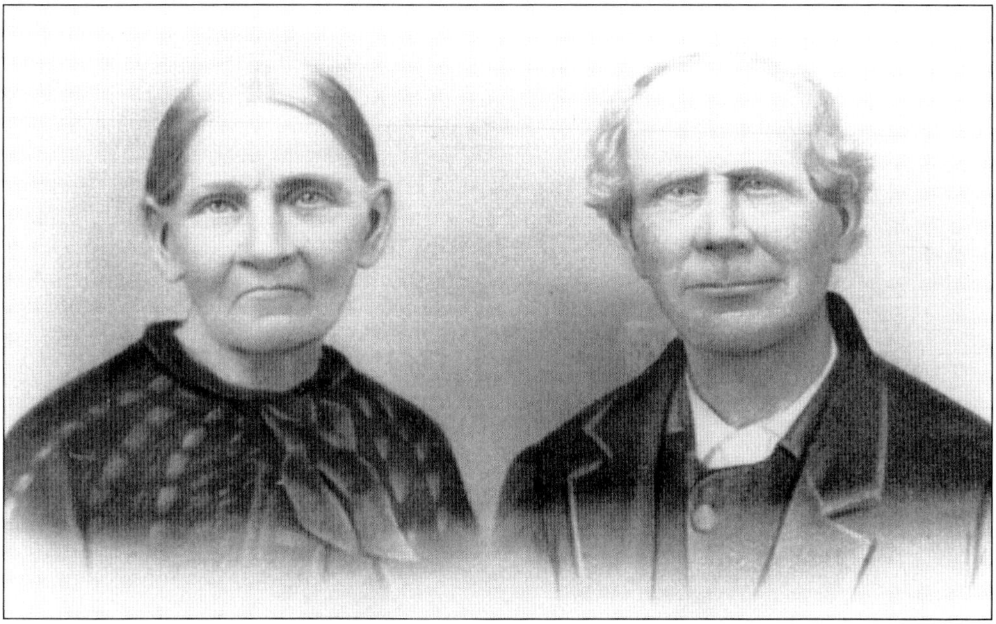

Nicholas Horn owned land southwest of Tremont in the area that is today known as Horn's Crossing in Bull Mountain Bottom. In the late 1830s, Nicholas moved to Itawamba County, where he married Jane Hill in 1841. Nicholas and Jane lost their eldest son, Jim, in 1864 in the Battle of Atlanta, the same year that their son Gaines Horne was born. (Courtesy of Patsy Adcock Kennedy.)

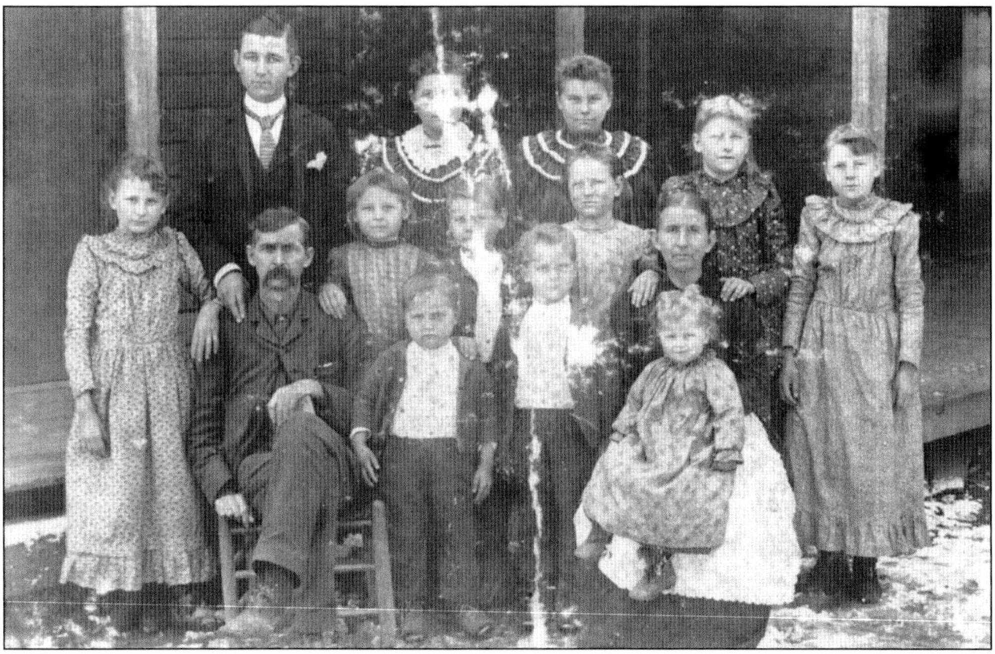

Benjamin Crittendon Mitchell and his wife, Nancy Ratliff, raised a large family in the Ratliff community. Their children were Tamsie Ophelia, Rozzie Ann, Verna, Bettie, Mary Alice, Harvey, George, William, Jimmie Dellar, Circie, Lillian, Mattie Belle, and Shaler Allen. (Courtesy of Coleen Thornton.)

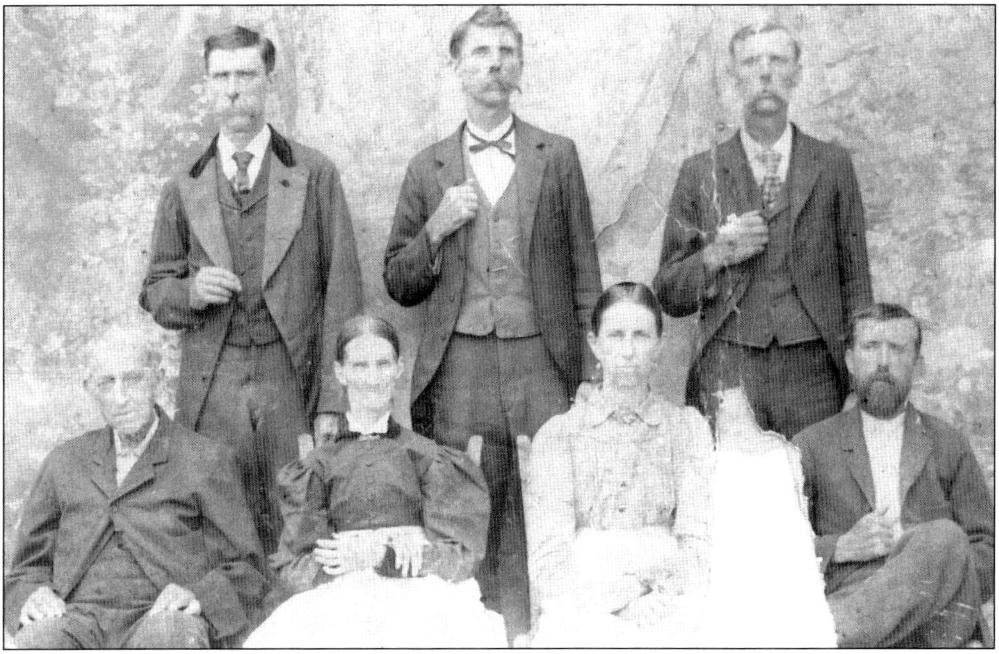

Henry Jackson Burch came to Itawamba County from Georgia in 1844. Following him, against her father's wishes, was Phoebe Webb Clark, whom he married later that same year. Henry and Phoebe raised their seven children in what is now the Clay community. Pictured from left to right are (first row) Henry Jackson Burch, Phoebe, Mary Frances, and Thomas Burch; (second row) John Henry, Allen, and George. Not pictured are William and Charles.

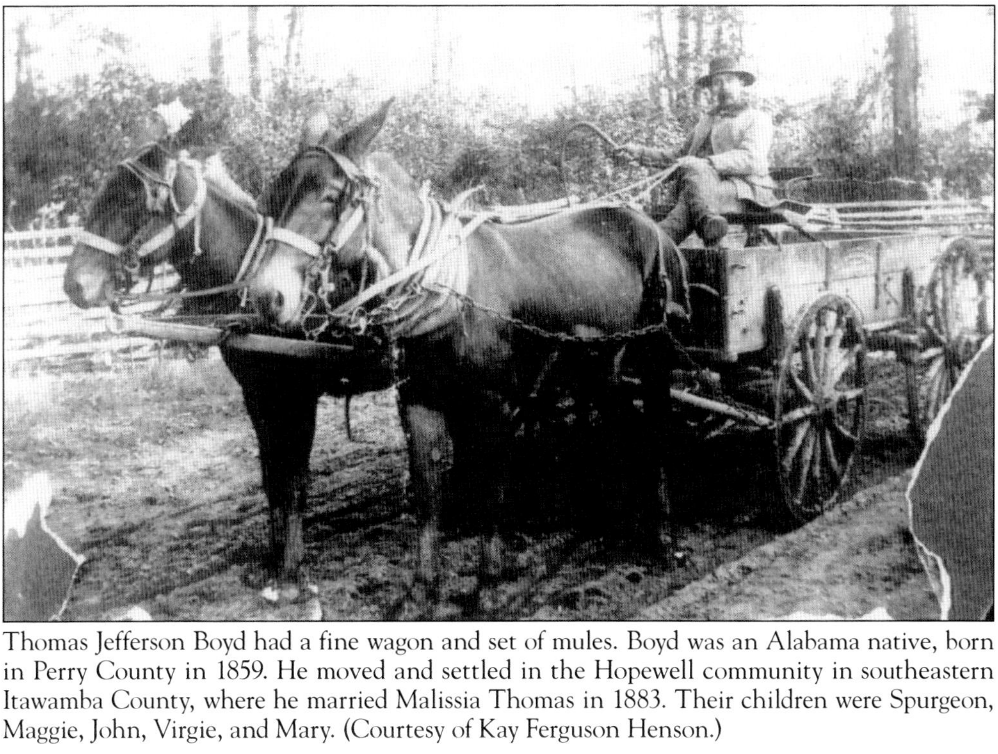

Thomas Jefferson Boyd had a fine wagon and set of mules. Boyd was an Alabama native, born in Perry County in 1859. He moved and settled in the Hopewell community in southeastern Itawamba County, where he married Malissia Thomas in 1883. Their children were Spurgeon, Maggie, John, Virgie, and Mary. (Courtesy of Kay Ferguson Henson.)

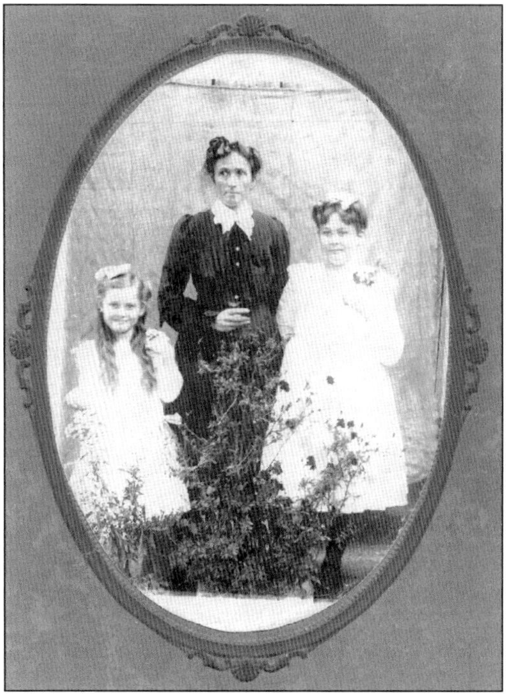

Traveling female photographers were not a common sight in rural Itawamba County; however, in this photograph from about 1900, an unknown female photographer poses with a couple of her young subjects. Lilly Evans Johnson, daughter of John Thomas Evans and Elizabeth Ann Bishop of the old Whitney community south of Tremont, is standing on the right while the photographer holds the shutter bulb in her hand.

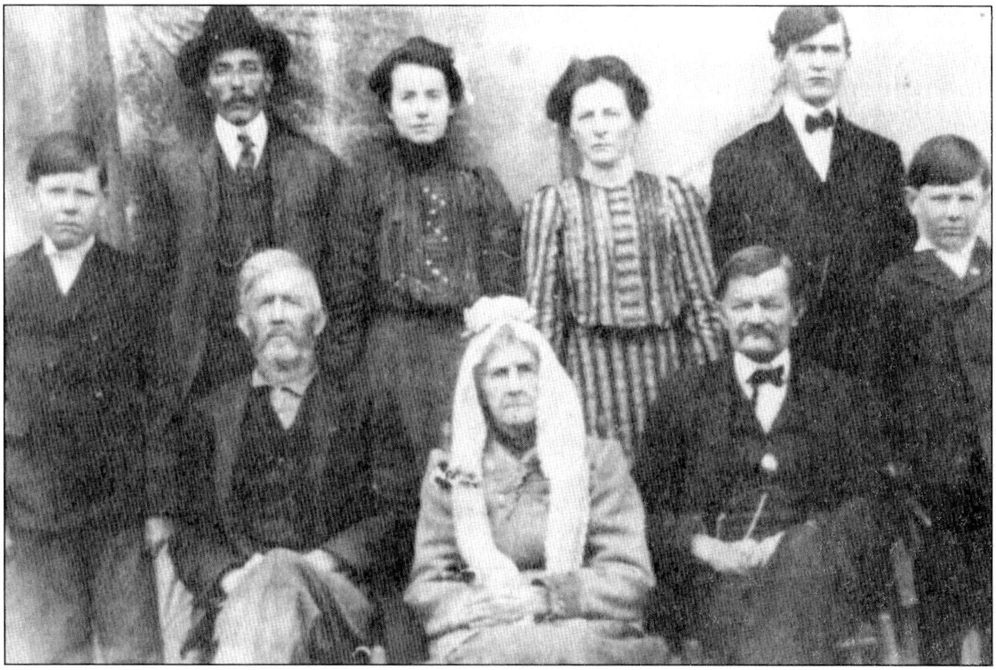

Henry T. Moxley, seated at left, lived with his family in the Bounds Crossroads area. Seated at far right is Henry's brother, Joseph D. Moxley, while at center is their mother, Emily Sims Moxley Sturkey. Standing, from left to right, are Henry's children—Richard, Florence with her husband John F. Williams, Emily, Jodie, and Luther. Two daughters, Alice and Mattie, not pictured here, married Dulaney siblings. The Moxleys and Williams were schoolteachers; Luther and Richard were Church of God ministers.

Born in 1883, Nervia May Dulaney Senter was the only daughter of Alfred "Babe" Dulaney and Lucinda Alabama "Allie" Chilcoat. May was widowed with nine children after her husband, Jesse Alvin Senter, died in 1924 from a small cut on his hand that led to blood poisoning. (Courtesy of Jessie Bane Senter, daughter of May and Jesse Alvin Senter, pictured here in her mother's lap.)

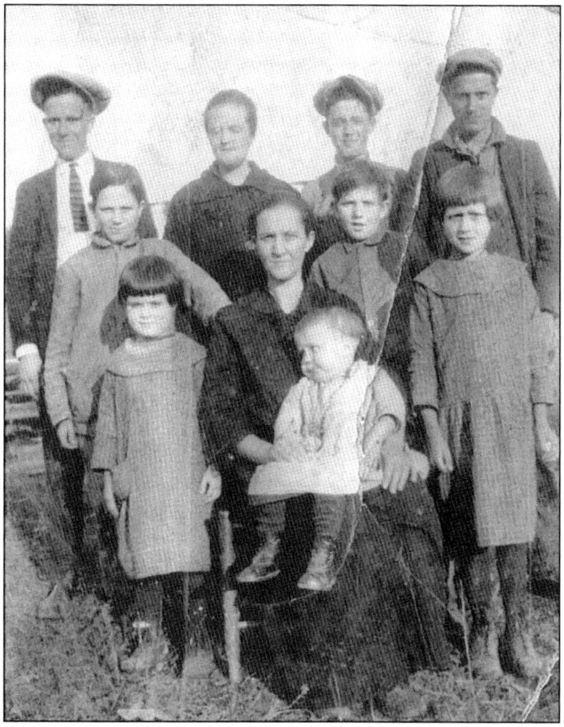

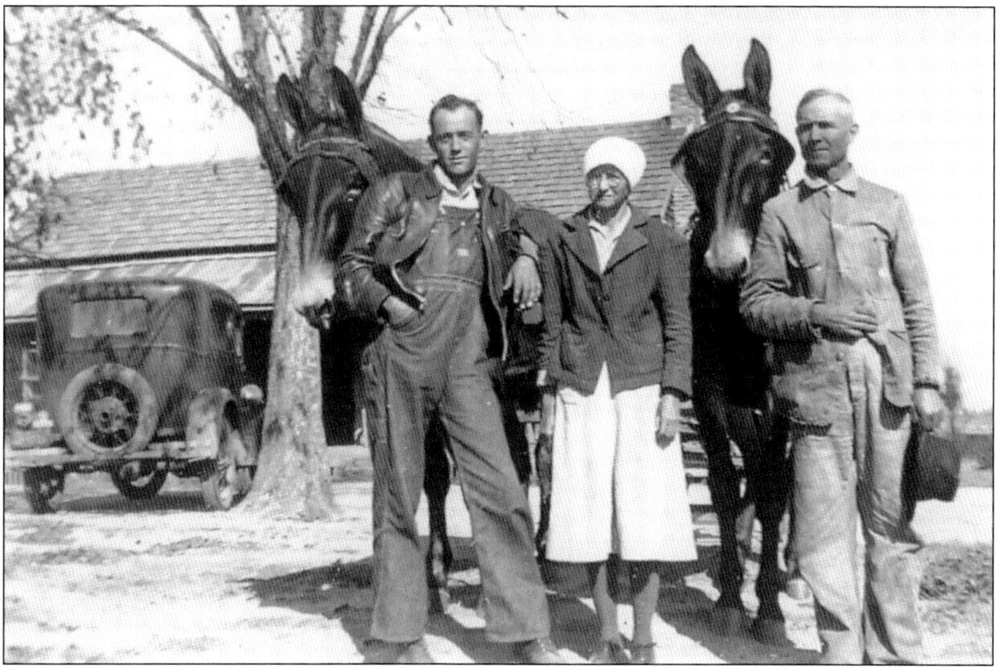

Mules were replaced in the field by tractors and on the road by automobiles. As more families purchased cars, mules were no longer featured in family photographs, replaced by the family's new mode of transportation. Porter Gainsville Dulaney and his wife, Arvilla Johnson, are pictured with Olun, their son, along with their mules *and* their car in this family photograph. (Courtesy of Terry Wilemon.)

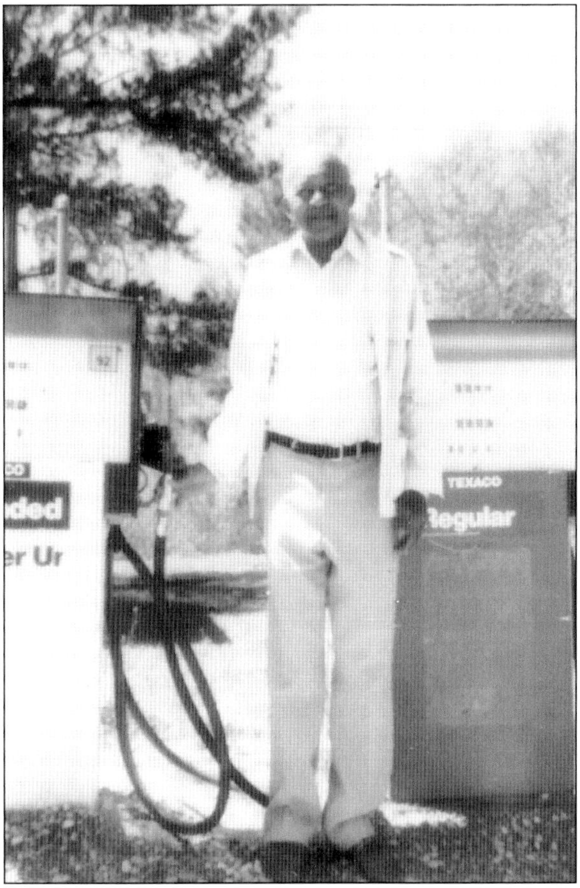

O.Z. Cummings, son of John and Jane Cummings, operated Cummings Grocery in Evergreen for many years. The store was a gathering place for the community. In addition to selling merchandise and gasoline, the store at various times included a barbershop, café, and pool hall. At one time, piano lessons were even taught in the store by Victoria Cummings. (Courtesy of Ormella Cummings.)

Charley Donald of the Dorsey community shows off the results of his trapping activities in 1955. Gaither Brothers Store bought the animals to resell to fur buyers such as Holley Brothers of Booneville. Donald's truck, parked on the court square, appears to be covered with raccoons, otters, and a couple of beavers. (Courtesy of the Delmus Harden family.)

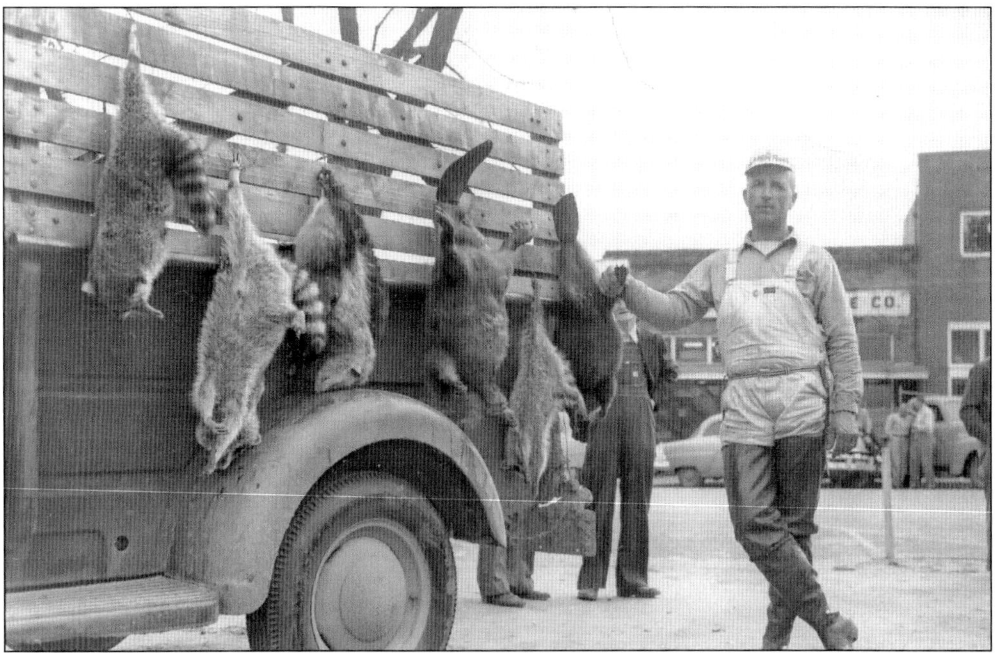

Mothers and their children gather on the front porch in 1938. They are, from left to right, (first row) Margie Mills, Ruby Mills, and Glader Johnson Mills holding children Vera Mae and Paul; (second row) Etoye Johnson, Dorothy Johnson Taylor holding her son Gene, and Syble Johnson Mills. Ruby and Margie were daughters of Syble; Glader, Dorothy, Etoye, and Syble were sisters.

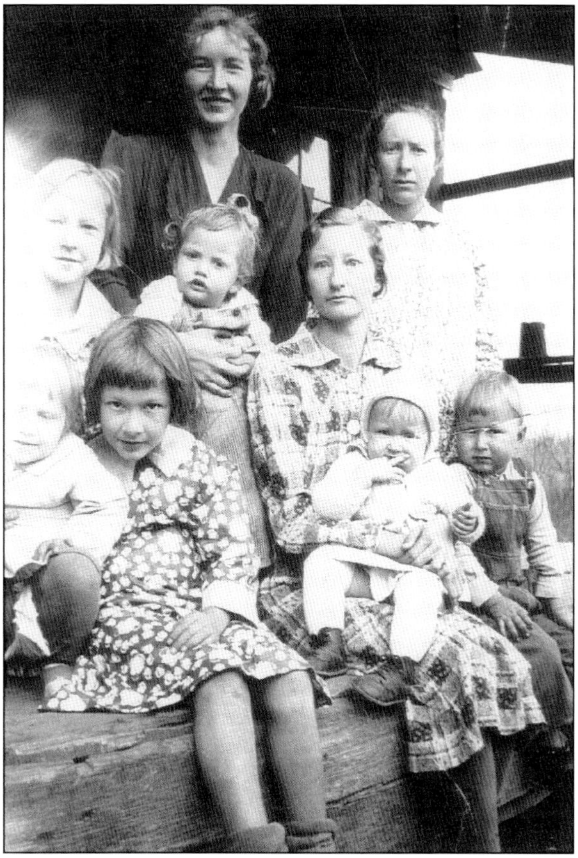

Clodie Casey, standing at far left, coached the Fulton Bars basketball team in the 1940s. Members of the team are, from left to right, (first row) Harold Clifton, L.C. Stone, Raymond Clifton, and Robert Hooper; (second row) Fred Stone, Isom Stone, J.D. Price, and Theodis Brownly. (Courtesy of the Delmus Harden family.)

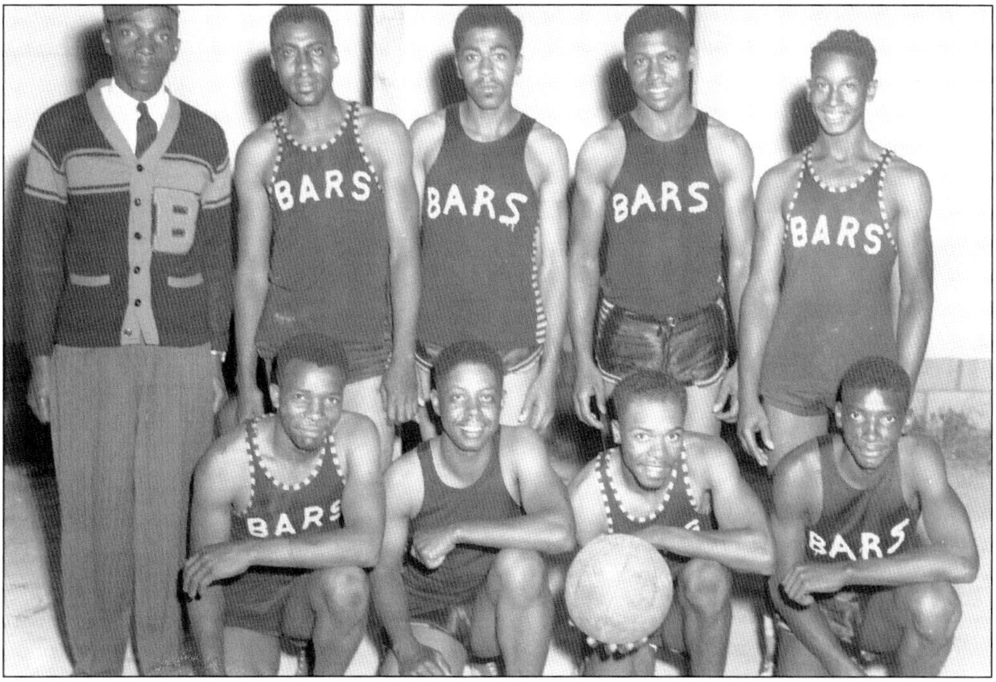

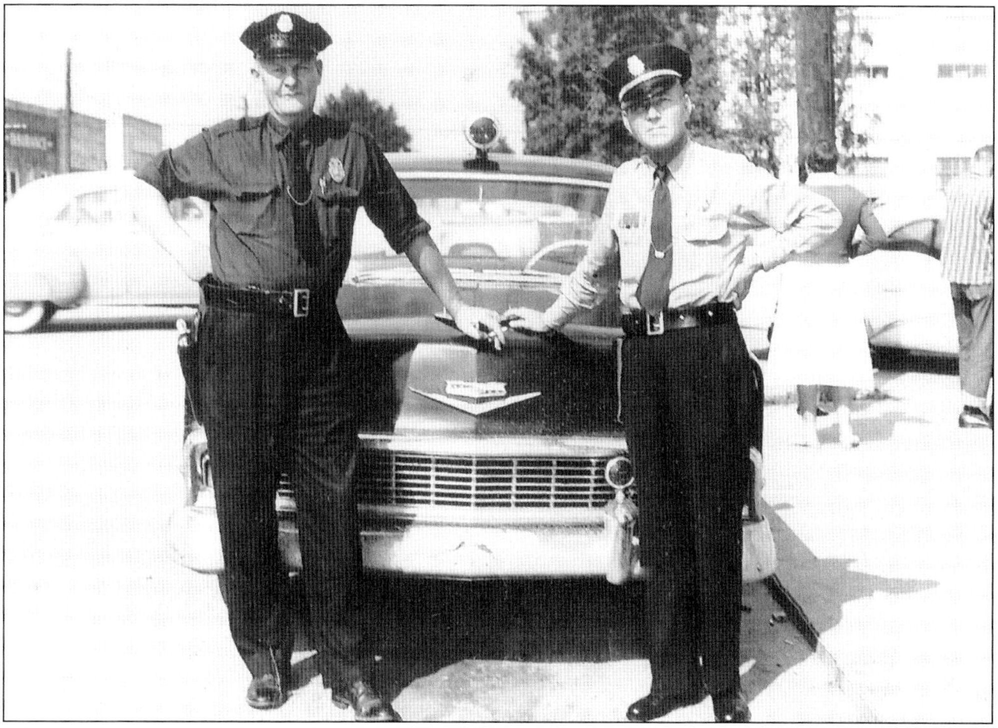

Edris L. "Cub" Dove, left, was the night marshal for the town of Fulton in 1957. J.O. Shaw, standing at right, was the day marshal. There being no chief of police at the time, these two men served as the local law enforcement for Fulton. (Courtesy of Johnny Taylor.)

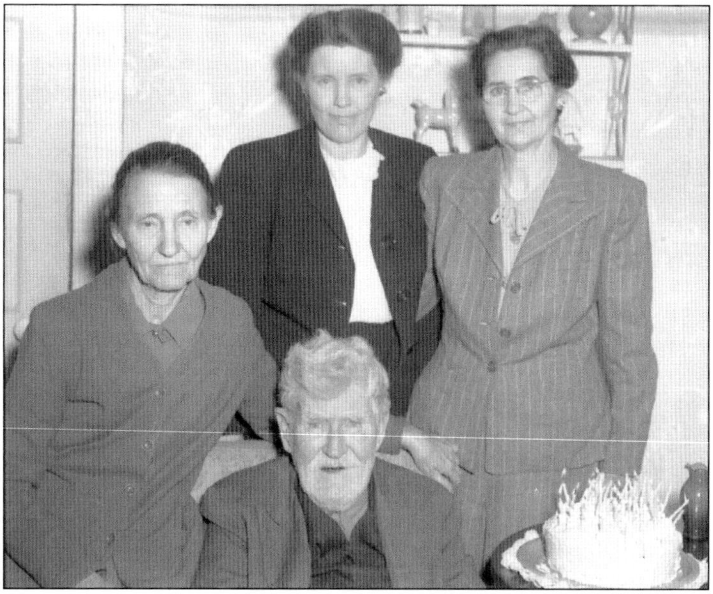

Joseph Elihu Sandlin taught in the county schools for over 40 years and served two terms as superintendent of education. He dedicated nearly half a century to the education of Itawamba County's children and taught at Sulphur Springs School when he was 80 years old. Professor Sandlin died in 1953 at the age of 93. He is pictured here on his 90th birthday with his daughters, Ethel (left) and Lilly (center), and his second wife, Mary.

A well-loved Itawambian, Zereda Greene preserved much of the county's history in her weekly columns for the *Itawamba County Times*. She was a teacher and principal, court clerk, civil servant, and faithful member of Fulton Methodist Church. Miss Zereda, as she was known, was born in 1897 to James Greene, a coffin maker and blacksmith, and Martha Jane Whitesides, owner of a millinery shop in Fulton. (Courtesy of the Delmus Harden family.)

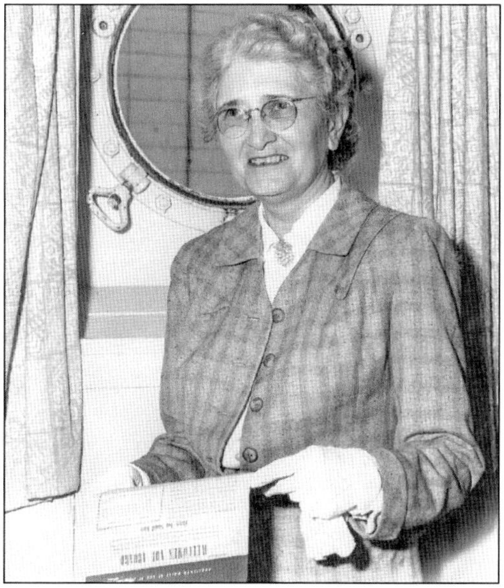

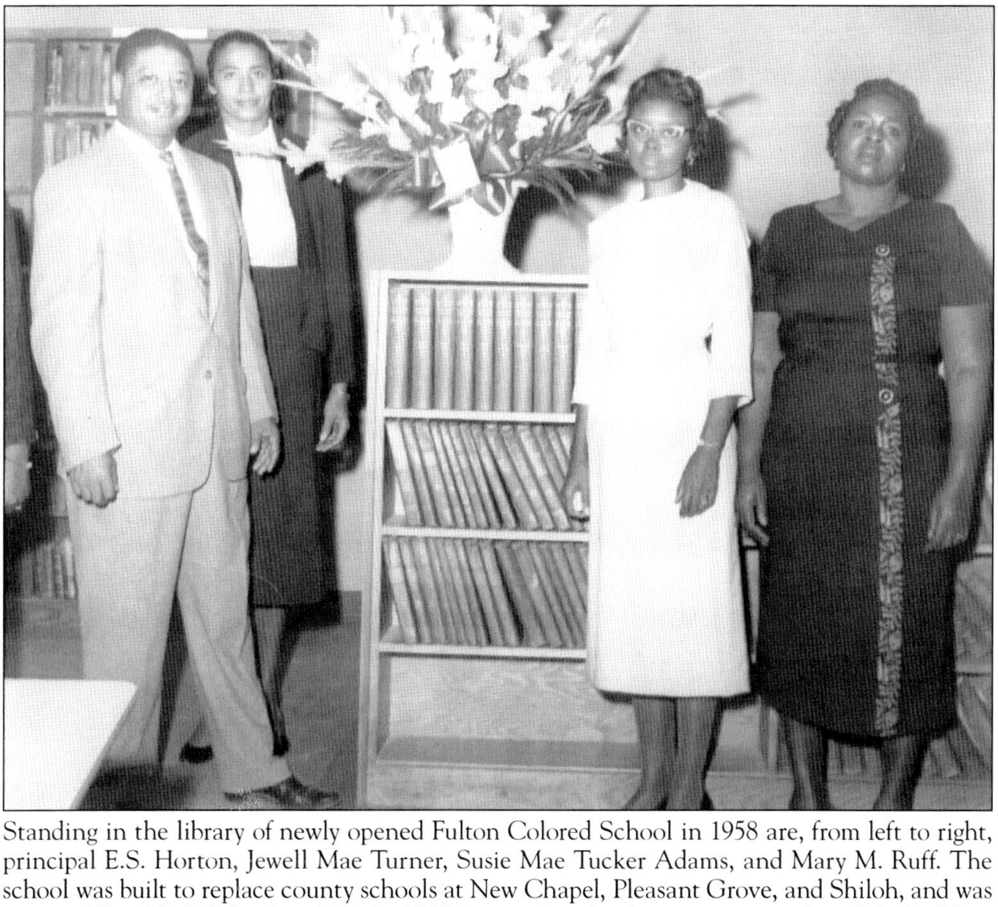

Standing in the library of newly opened Fulton Colored School in 1958 are, from left to right, principal E.S. Horton, Jewell Mae Turner, Susie Mae Tucker Adams, and Mary M. Ruff. The school was built to replace county schools at New Chapel, Pleasant Grove, and Shiloh, and was renamed East High School in 1962. In 1968, Itawamba schools were integrated, and the building that housed East High was abandoned as a school. (Courtesy of the Delmus Harden family.)

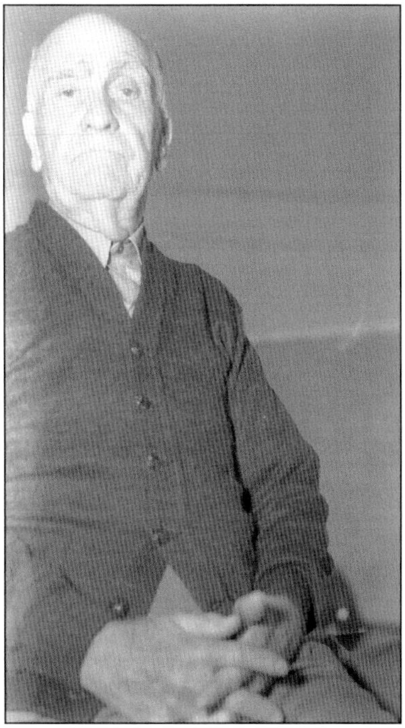

Alla Dorus Graham partnered with Dr. J.M. Walker in 1900 to publish the *Fulton Herald*. For the next 35 years, Graham was involved in the newspaper business in Fulton, editing and/or publishing the *Fulton Herald*, *Itawamba County News*, and *Fulton News Beacon*. Graham also served several terms as mayor of Fulton. A Fairview native, he was married to Verdie Clifton, and they had one daughter, Hermine Graham. (Courtesy of the Delmus Harden family.)

Delmus C. Harden purchased the *Itawamba County Times* in 1945 and served as its publisher until his death in 1977. Harden epitomized the community newspaper editor. He realized the success of the newspaper depended upon the success of the community, and vice versa. Harden was instrumental in the establishment of Itawamba Junior College, the first community hospital in the county, Fulton's first fire department, and many other enterprises. His legacy lives on, not only in his photographs that appear in this book, but also in the contributions he made to the success of the county.

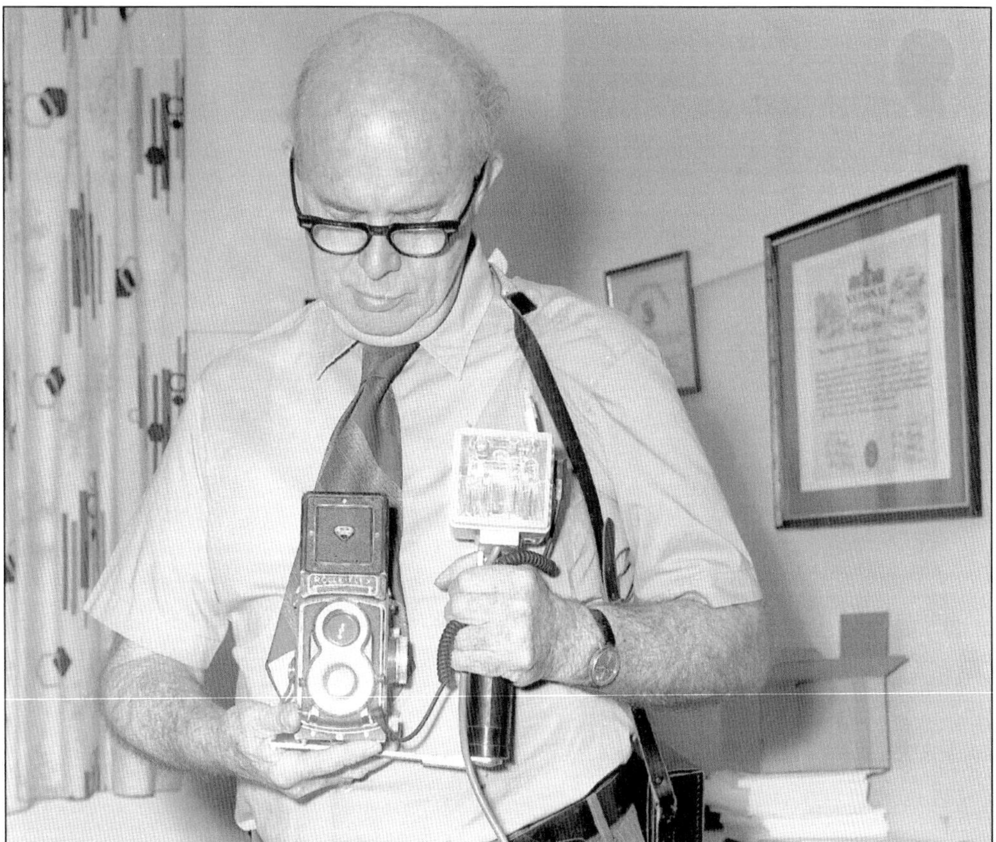

Five

HOORAH FOR ITAWAMBA!

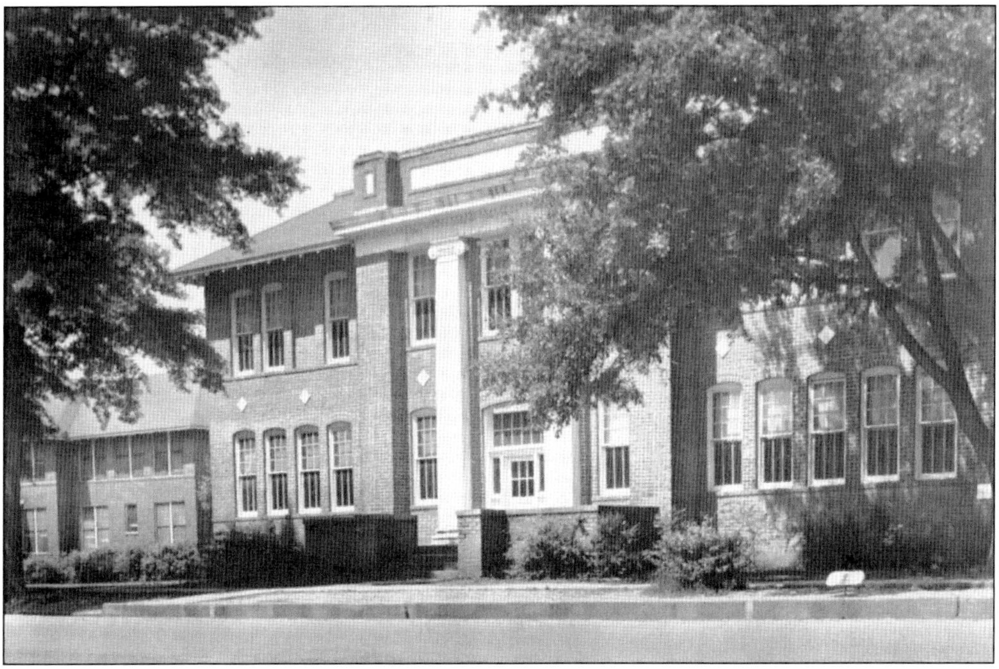

Itawamba Agricultural High School (IAHS) opened its doors in September 1921 on a campus west of Fulton along Bankhead Highway, which also was completed that same fall. In 1948, Itawamba Junior College began classes, coexisting on the same campus until IAHS moved south of Fulton in 1993. Throughout their nearly 50 years together, the two institutions shared common buildings (including the Administration Building shown here), school annuals, bands, athletic facilities, and even instructors. Their histories are closely intertwined.

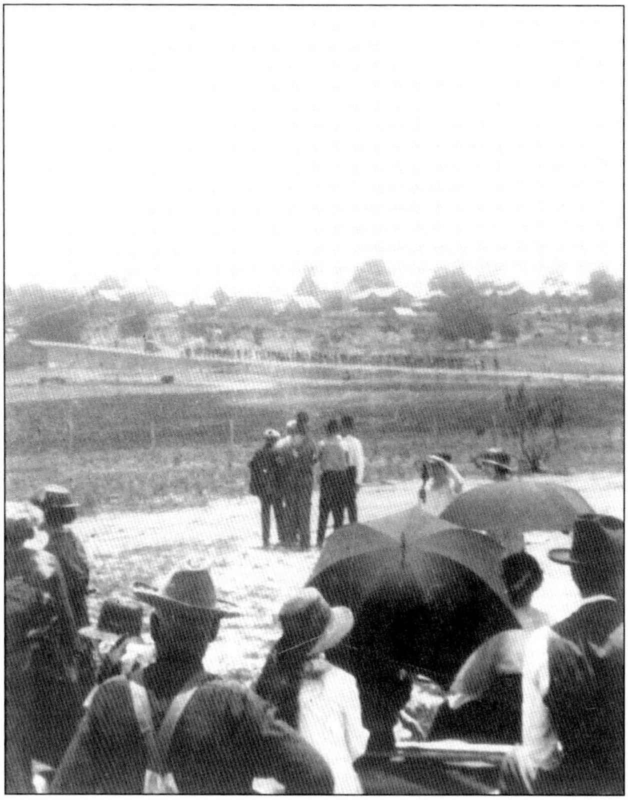

Citizens and dignitaries parade down the hill from the court square in Fulton prior to the laying of the cornerstone of the Administration Building that would serve the county high school and later the college. In 1908, the Mississippi legislature passed a bill that provided for the establishment in rural counties of agricultural high schools that stressed agricultural education for boys and home economics for girls, in addition to a regular curriculum. Legislation in 1922 allowed qualified agricultural high schools to offer college classes, and in 1929 the possibility of adding a junior college on the IAHS campus was considered. The Great Depression and World War II delayed any such plans, but efforts were renewed in 1946, resulting in the opening of Itawamba Junior College in 1948.

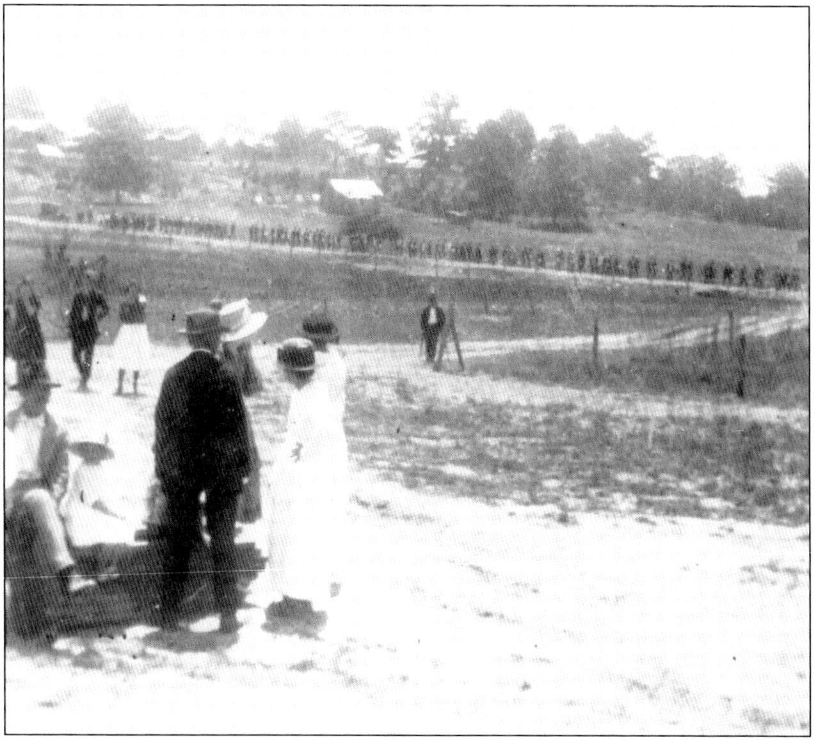

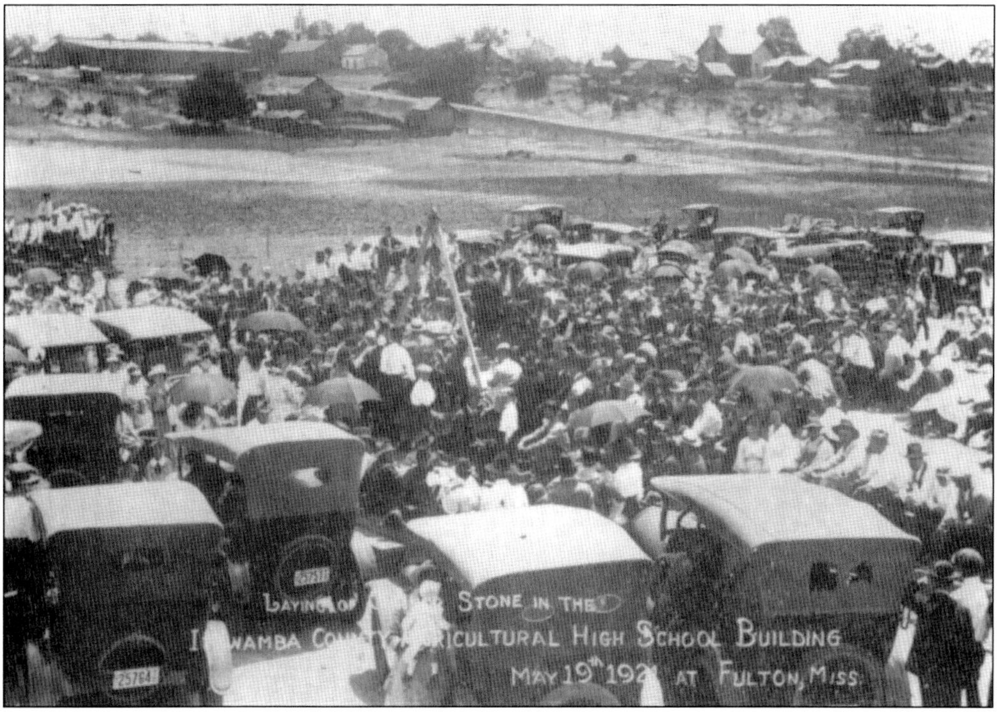

The laying of the cornerstone for the first four-year high school in Itawamba County took place on May 19, 1921. The cornerstone was put in place during a ceremony conducted by members of the Masonic lodge and contained a Bible, a statement of the Fulton Bank, and a copy of the *Itawamba County News*. It was located on the northeastern corner of the Administration Building, visible in later pictures of that structure.

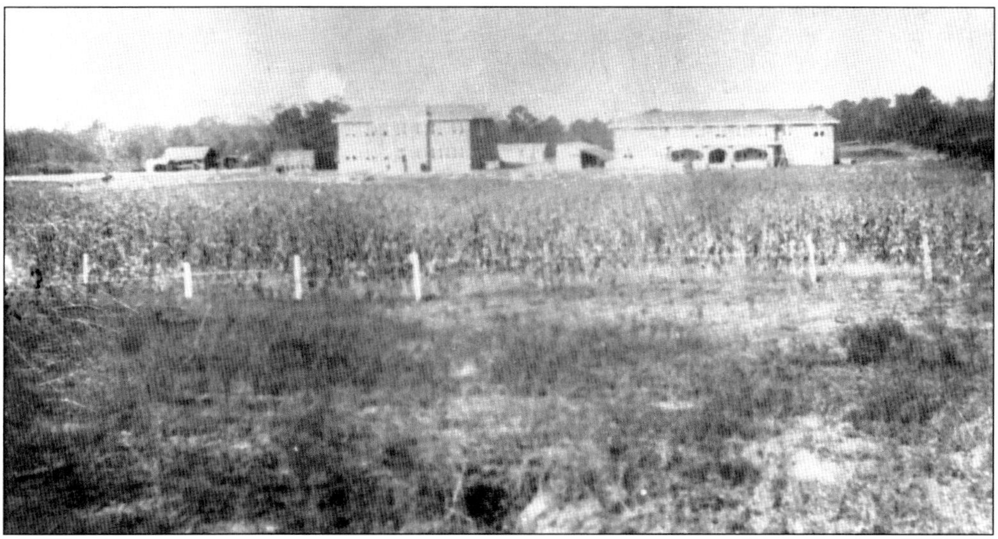

One of the earliest photographs of the high school campus, taken before the boys' dormitory was built late in 1921, shows the Administration Building (center) and girls' dormitory (right). The building to the far left rear was the school's barn. IAHS had nearly 100 acres of land and was somewhat self-sufficient; students were required to provide a few hours of labor each week. Today, the college's administrative offices are located where the girls' dormitory once stood.

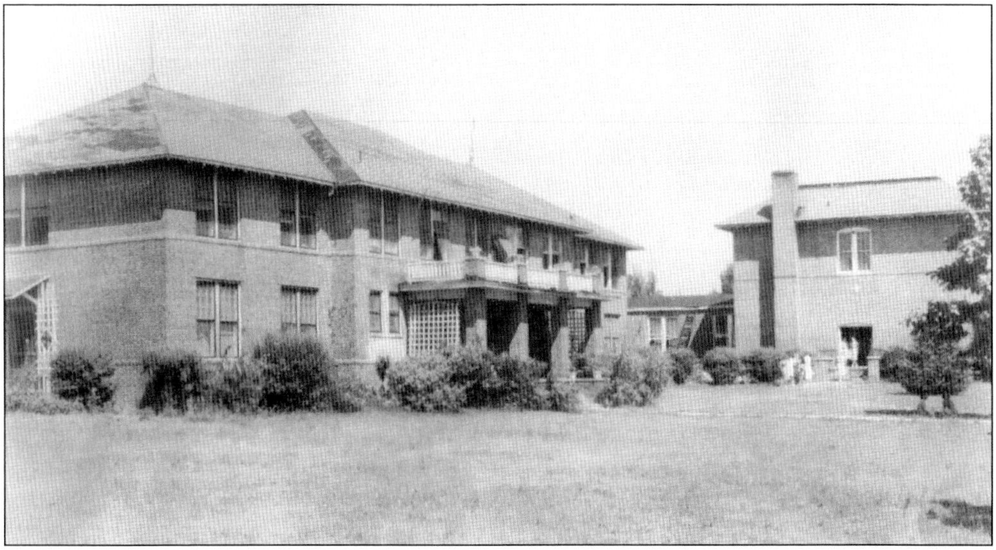

The boys' dormitory was originally a wood building called the "Bull Pen" by students. Built in 1921, the wooden structure burned in 1927 and was replaced by the brick building shown here, located on the south side of the Administration Building. The dormitory became known as Lee Hall upon organization of the college. It was torn down in 1964 to be replaced by the college's Business Education Building.

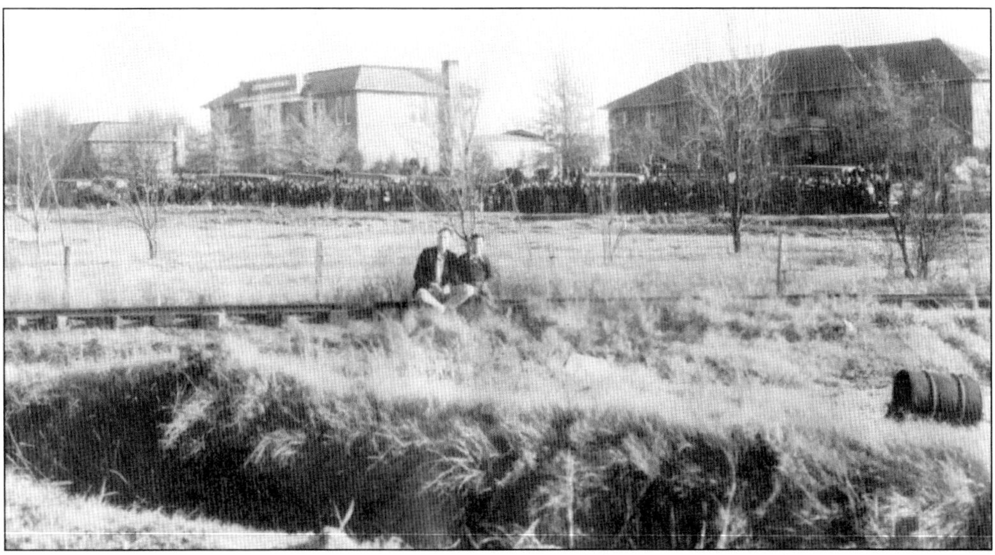

In this 1939 scene, buses line up at the end of the school day to return students to their homes. When the high school first opened, most students boarded on campus due to poor roads and inadequate transportation. From left to right are the boys' dormitory, Administration Building, and girls' dormitory. The Administration Building housed classrooms and an auditorium in addition to the school's offices.

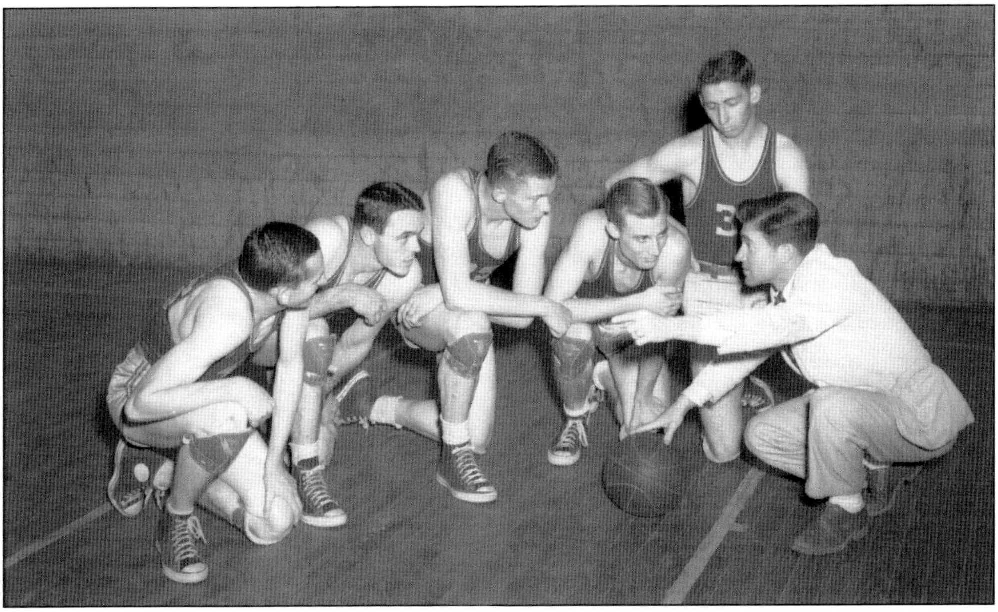

Even though the college had just started its first semester in the fall of 1948, athletic activities were immediately recognized as being an important part of campus life. Players from the 1948 boys' basketball team, coached by E.R. "Red" Garafalo, are, from left to right, two unidentified, Eddie Cayson, Dot Leathers, and Kermit McFerrin. The team shared a gymnasium with high school students, the same building that Itawamba County primary schools used for their tournaments. A brick gymnasium built in 1950 was torn down in 2012. Below, the main entrance to the campus in 1956 shows the Administration Building on the left and the structure that housed the high school's classrooms and auditorium on the right. Today, this street is a pedestrian walkway. (Both, courtesy of the Delmus Harden family.)

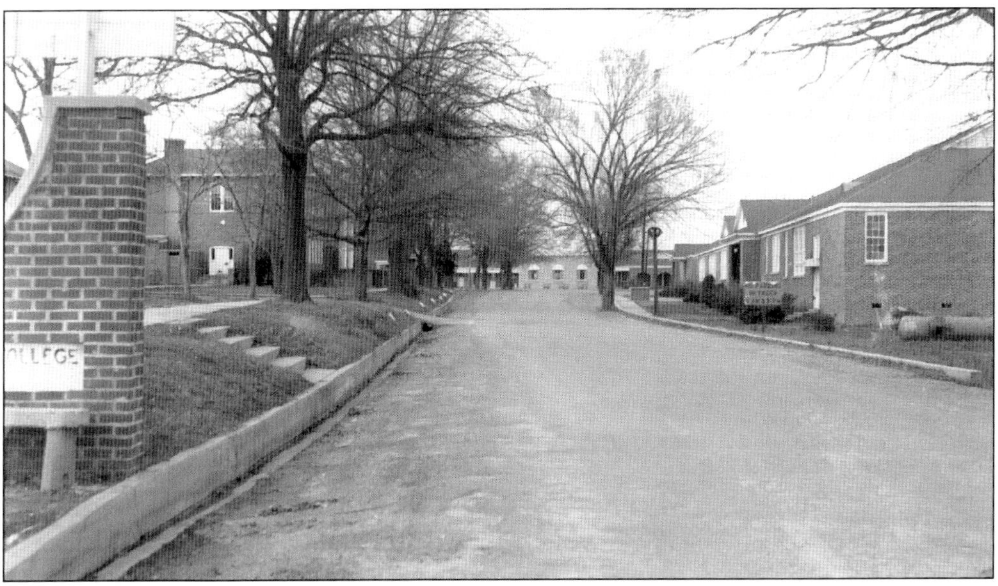

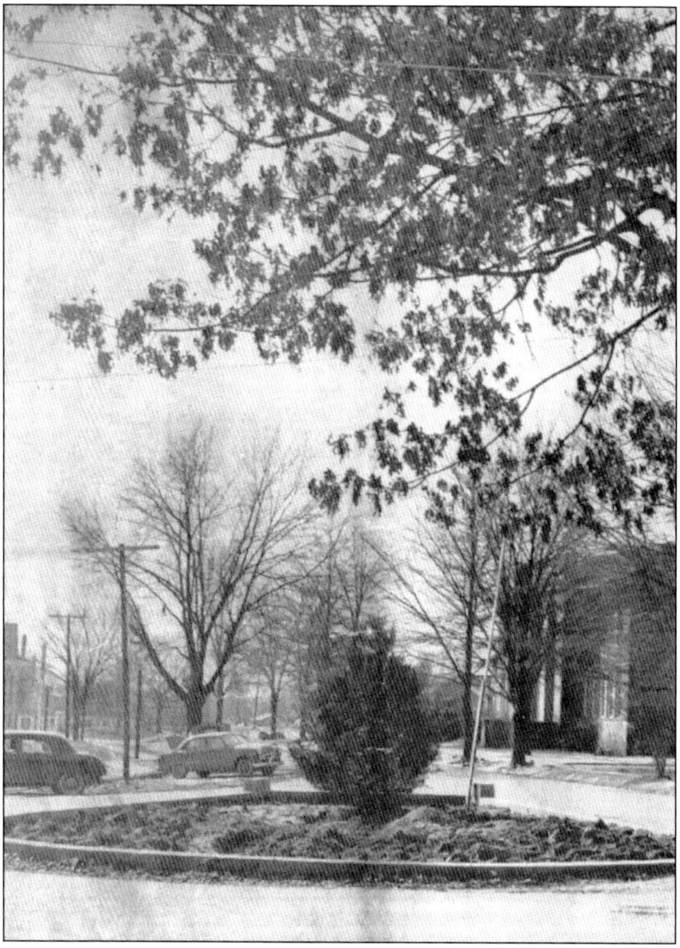

Located at the northern end of the original, primary street of the campus, the Horseshoe served as a turnaround spot for traffic. The college's present-day pedestrian street served as the main traffic artery for years. The old Administration Building stands at the far right in this 1949 photograph; it served both the college and the high school until torn down in 1964 to make room for a new liberal arts building. The Horseshoe was also a popular gathering spot for students waiting on class or a bus. Visiting with each other in the 1954 scene below are, from left to right, unidentified, Carolyn Burch, Shirley Repult, and Jimmy West. Behind them is the Recreation Hall, later to become the site of a cafeteria shared by both the high school and college. (Boh, courtesy of the Delmus Harden family.)

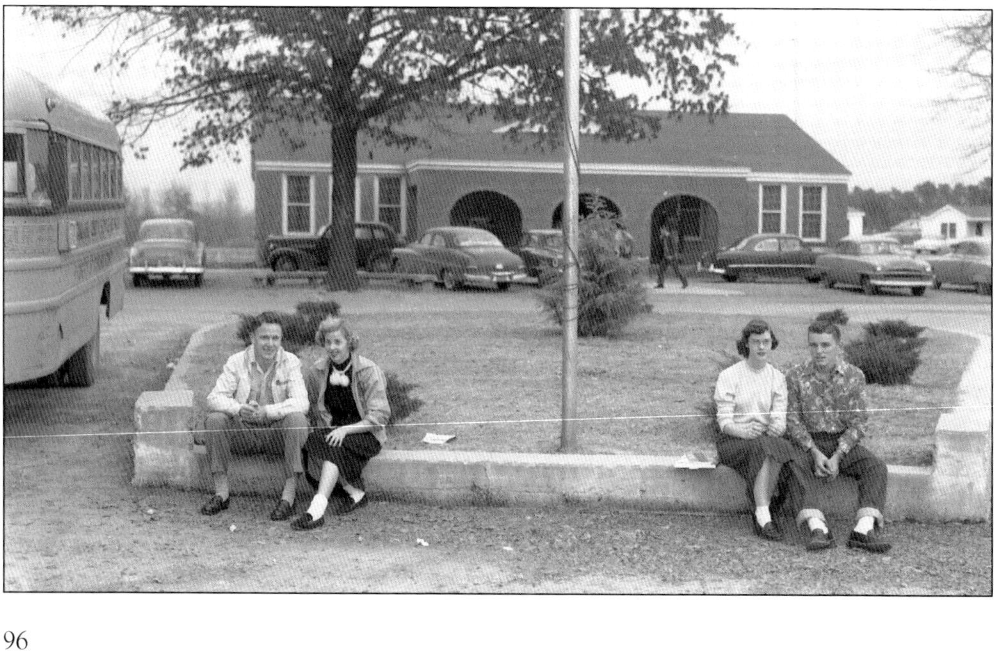

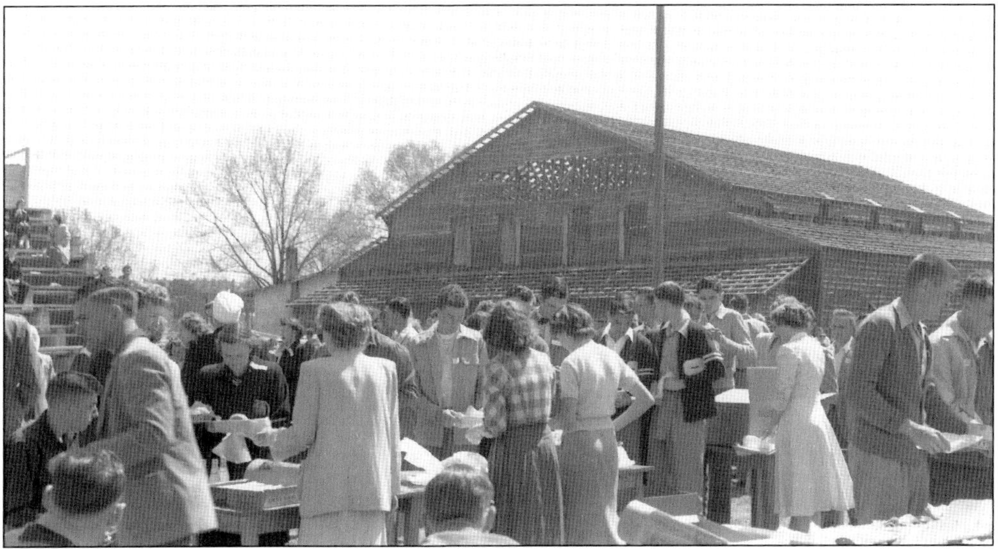

Students are being served lunch on the campus football field during field day activities in 1950. In the background, the old gymnasium is being dismantled, and, in the distance, the flag from the Horseshoe can be seen. Today, the Crane Pavilion is located on the site of the former football field and houses the first school bell used by the high school. Below, a mill and lumber company can be seen beyond the visitors' bleachers of the old football field. The college had over 100 acres in its possession and leased some of its unused acreage to the Gibbs, Haynes, and Kilpatrick mills. Today, the Health Science and Natural Science Buildings stand where the pictured lumber mill used to be. (Both, courtesy of the Delmus Harden family.)

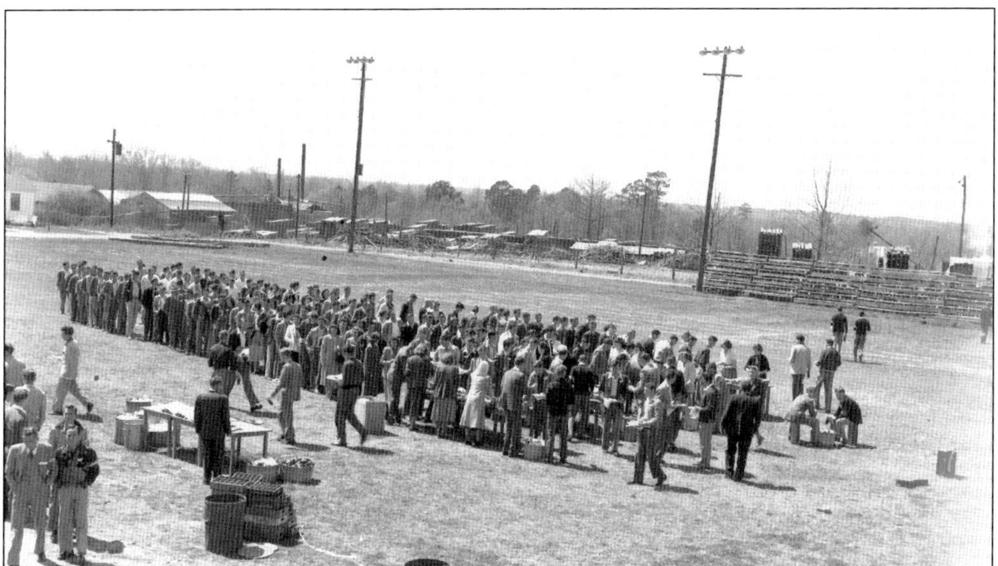

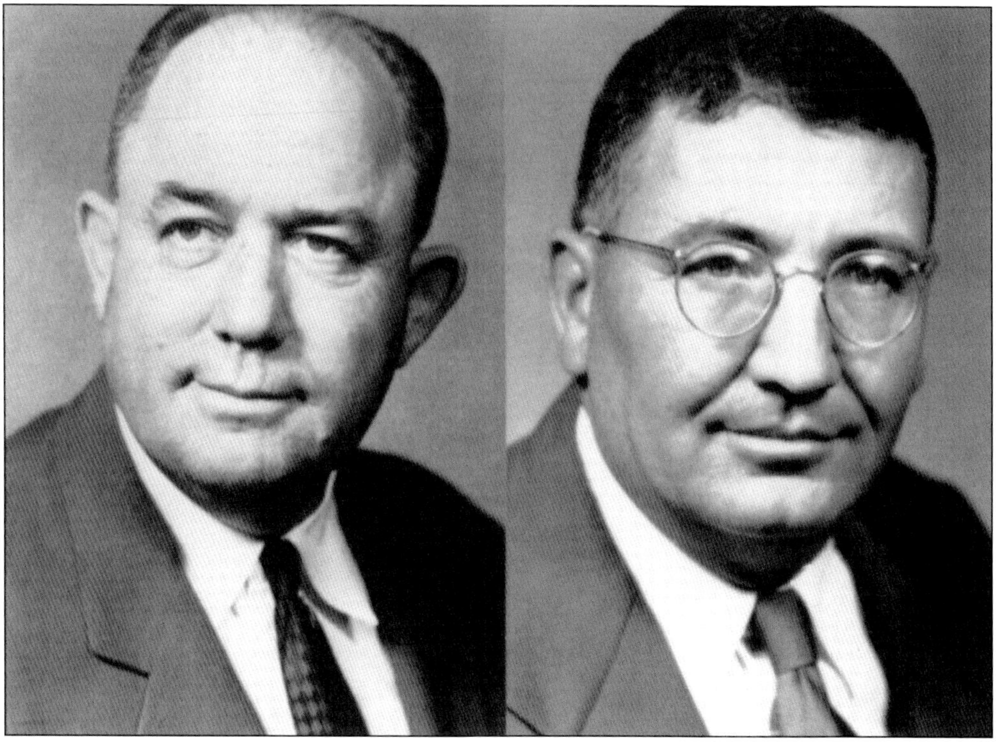

Phillip Sheffield (left), who was superintendent of IAHS in 1948, became the college's first president and served until his untimely death in 1960, at which time John S. Crubaugh, recruited by Sheffield as the first dean of students, took over. Both men have been credited with the early success of the community college. In 1997, four years after physical separation of the campuses, the administration of the high school was transferred to the county.

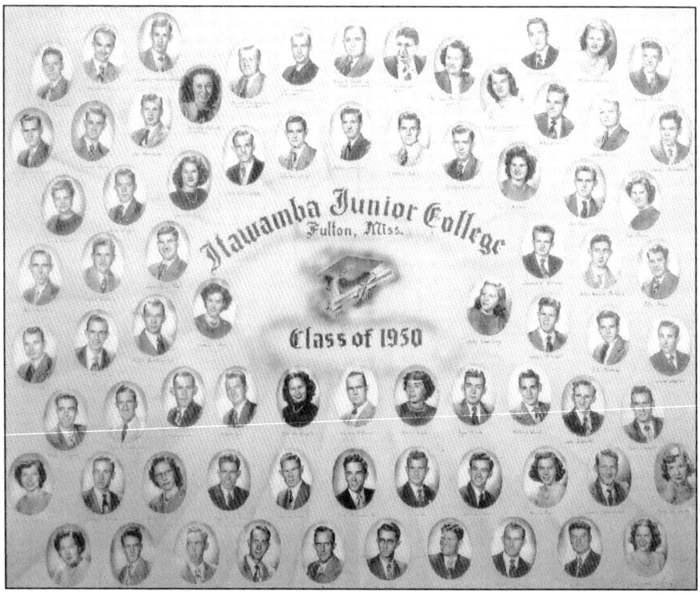

Over 70 students made up the first graduating class of Itawamba Junior College in 1950. When the college began in the fall of 1948, some 114 traditional academic students from Itawamba, Lee, and Monroe Counties registered for classes along with 170 veterans of World War II, who enrolled in vocational training. Full-time enrollment in fall 2011 was 5,014 students on the Fulton, Tupelo, and Belden campuses of Itawamba Community College.

Six

TEACHING AND PREACHING

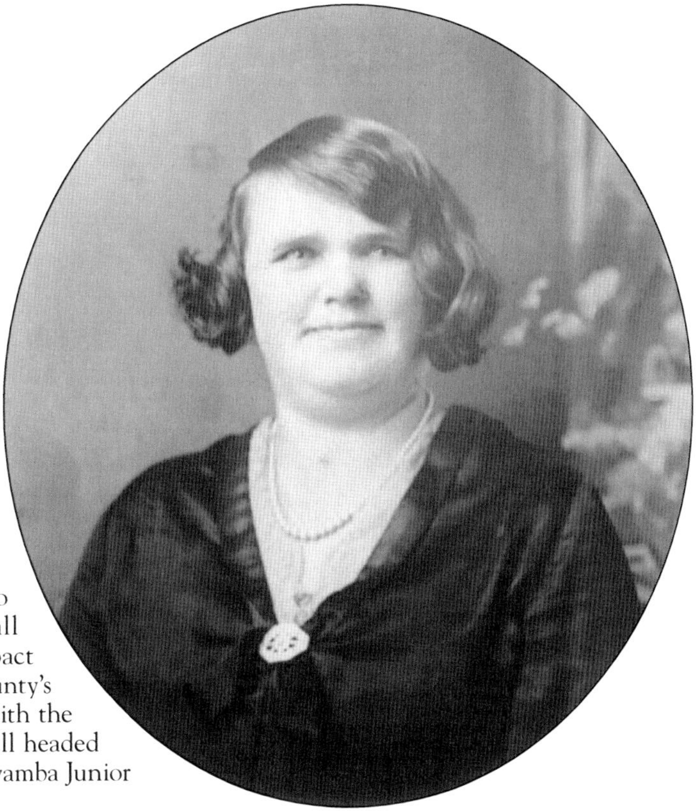

Elected in 1942, Delaphare Barkley Mayhall was Itawamba County's first female superintendent of education and one of the first teachers in the county to receive a master's degree. Her husband, Sam, served as superintendent from 1928 to 1932. Together, the Mayhall educators had a lasting impact on the education of the county's population. After a career with the county schools, Mrs. Mayhall headed the math department of Itawamba Junior College until 1966.

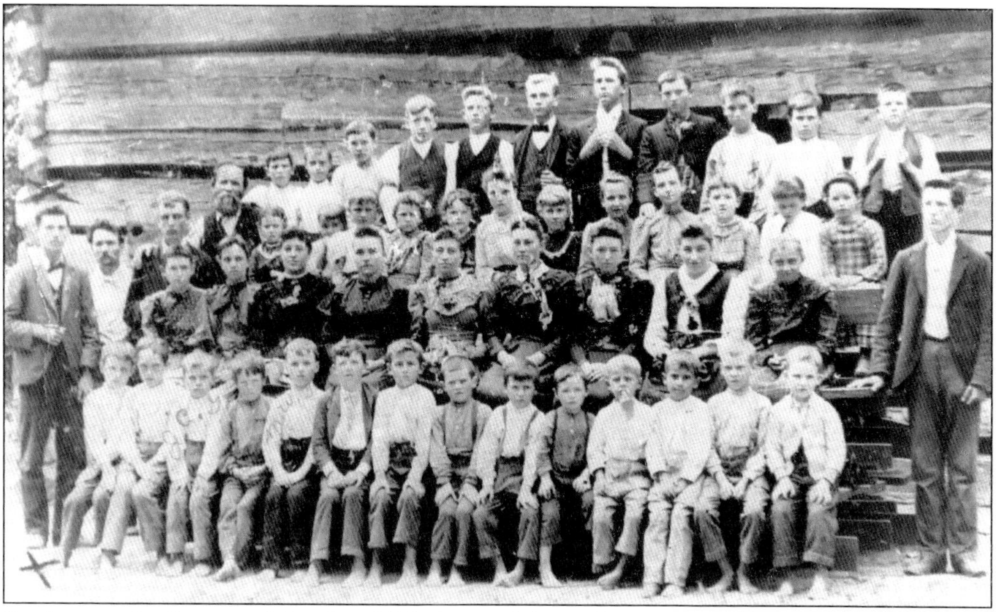

Students at the Fairview School in 1897 obtained their education in a log building, above, that was located in the area of present-day Fairview Baptist Church. Around 1900, a new two-story schoolhouse was built with the second floor used as a Masonic lodge and general meeting space. The large building was divided by a huge curtain down the center, with the lower grades taught on one side and fifth through eighth grades on the other. Older students paid tuition to attend higher-level classes. Below, principal and early Itawamba County educator Sam Mayhall is standing at far right; his fellow teacher was Ina Graham. Professor Mayhall began teaching at Fairview School in 1896; he taught in other schools throughout the county as well. He also served two terms as county treasurer and as superintendent of education.

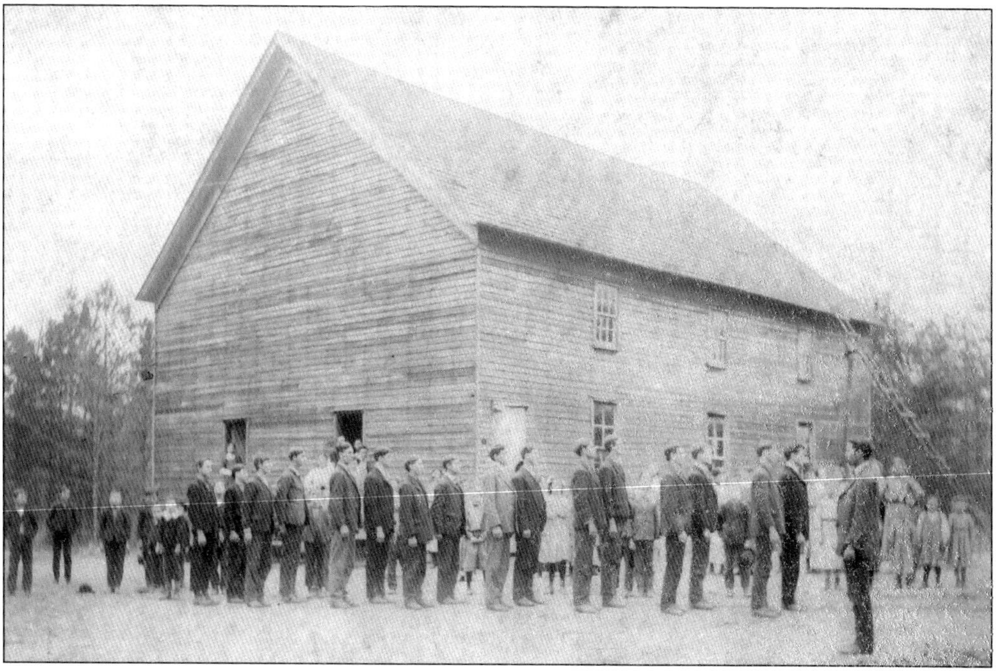

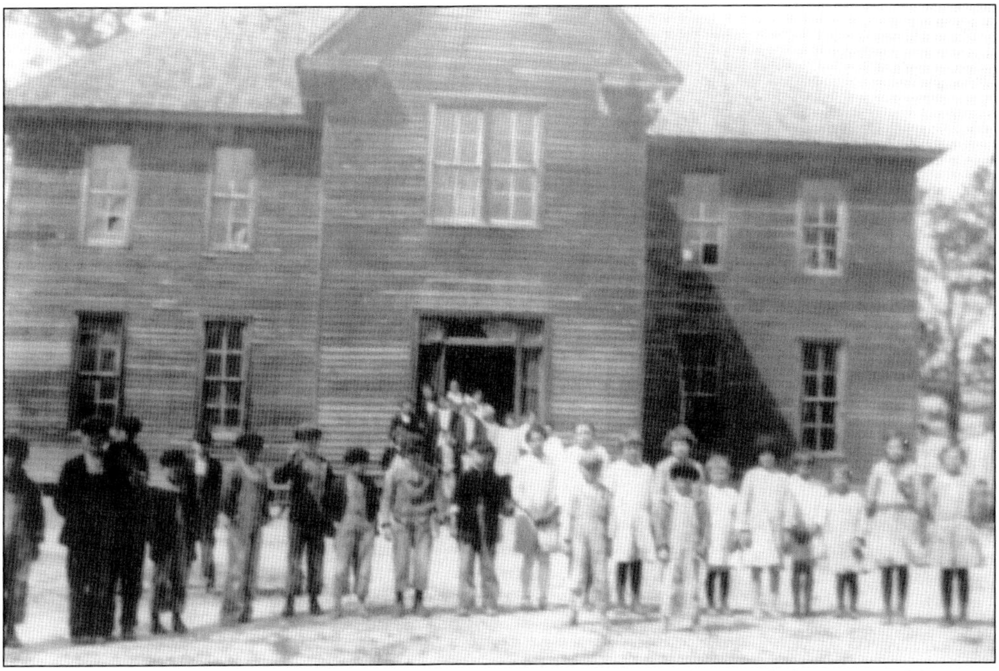

This two-story school building in Carolina, built in 1905, was damaged during a 1933 storm and was replaced with a brick structure. The people of the Carolina community were among the first in the county to supplement the free school term with an extra tuition term, and they later implemented a special levy for more teachers and a longer school term. (Courtesy of Francis Pennington Vann.)

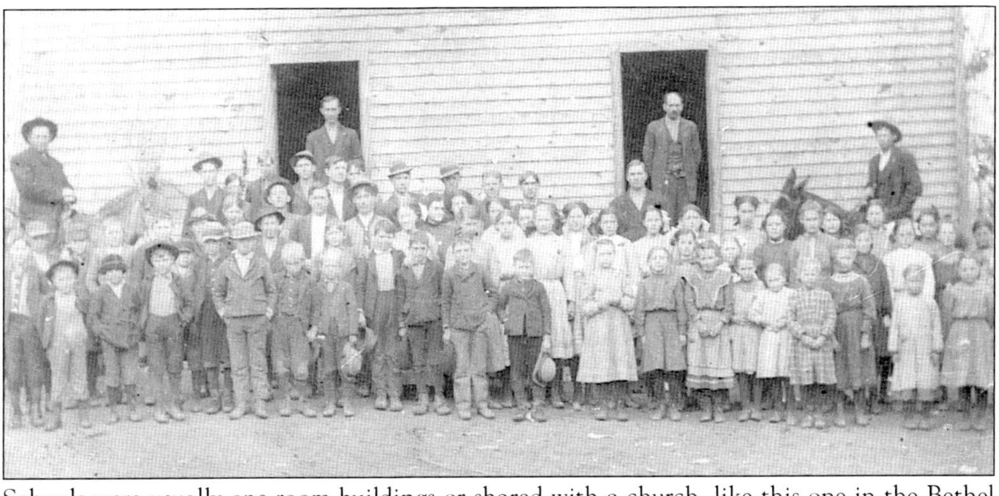

Schools were usually one-room buildings or shared with a church, like this one in the Bethel community in 1900, where Sidney Sheffield (standing in the door on the right) was a teacher. The earliest schools were primarily fee-based and took a hard hit following the Civil War. Around the 1880s, after Reconstruction, communities began to build back, and by the turn of the 20th century, more than 100 schools dotted the landscape in Itawamba County. (Courtesy of Joe Nell Wood.)

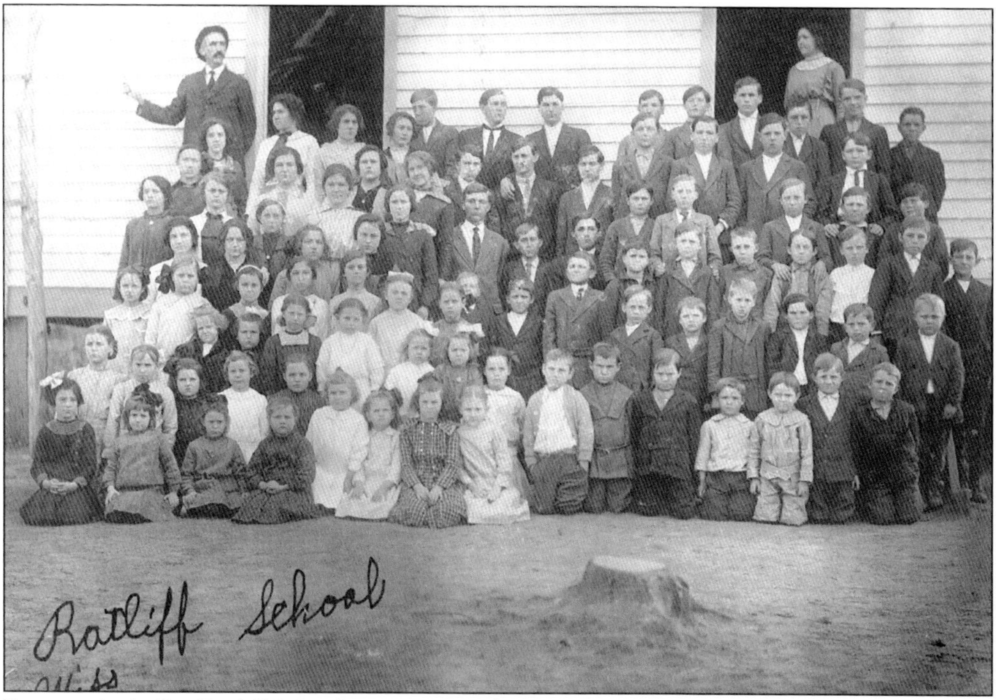

Ringing the bell at Ratliff School in 1913 is teacher C.A. Gilbert. Beatrice Gillespie, standing in the doorway, was a teacher that year as well. Ratliff School was closed in 1936. The number of county schools started dwindling when transportation other than by foot or wagon was possible. Population decline in rural areas also fueled the closure of community schools.

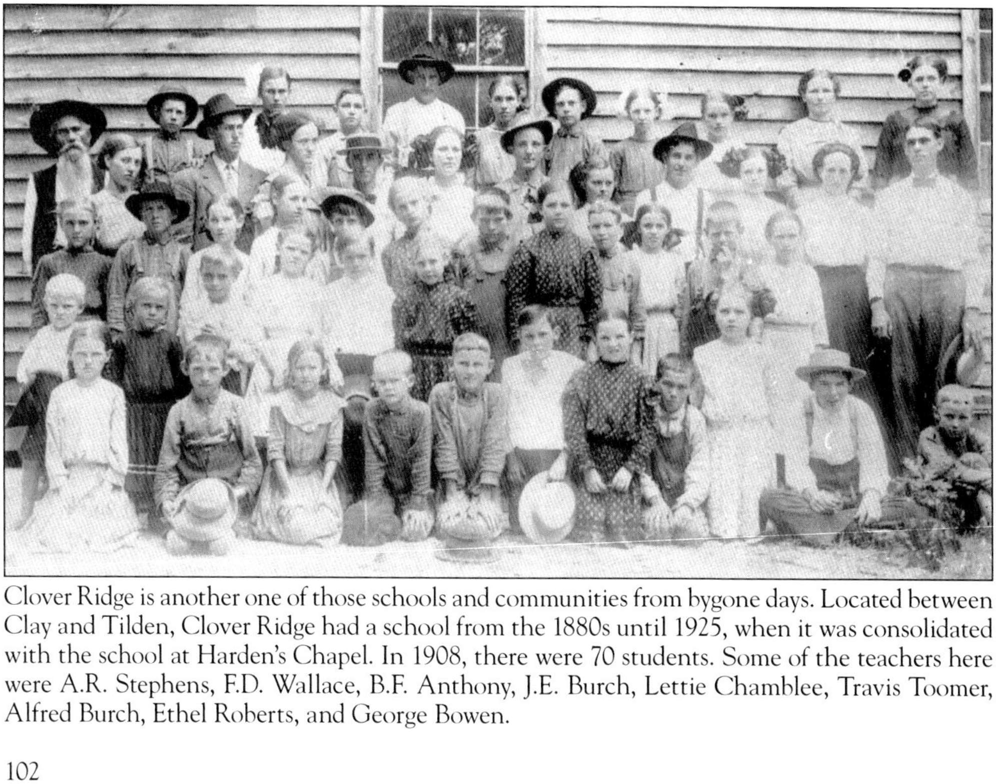

Clover Ridge is another one of those schools and communities from bygone days. Located between Clay and Tilden, Clover Ridge had a school from the 1880s until 1925, when it was consolidated with the school at Harden's Chapel. In 1908, there were 70 students. Some of the teachers here were A.R. Stephens, F.D. Wallace, B.F. Anthony, J.E. Burch, Lettie Chamblee, Travis Toomer, Alfred Burch, Ethel Roberts, and George Bowen.

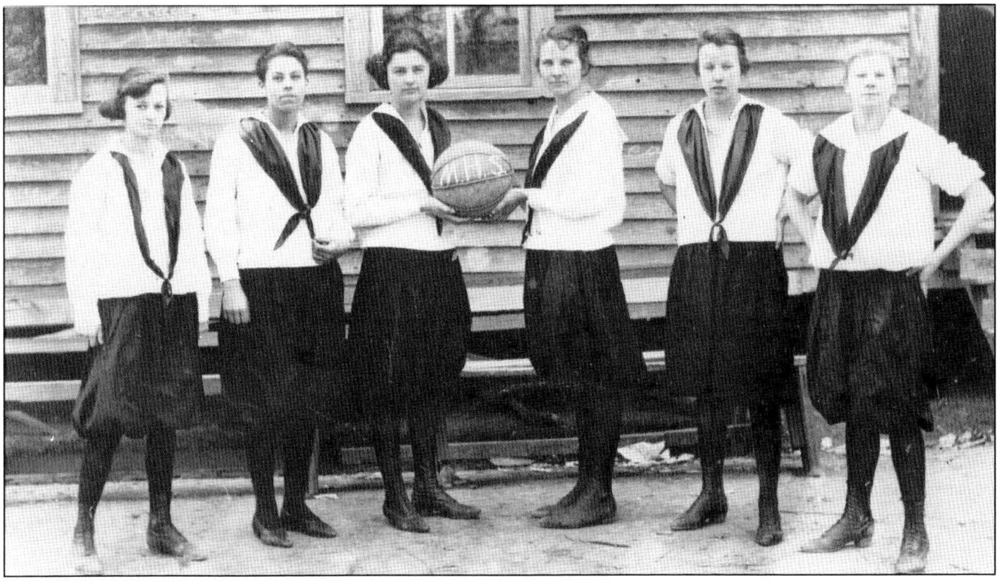

Athletic uniforms definitely have evolved over the years. This Mantachie School girls' basketball team from the early 1900s wears black stockings, lace-up boots, scarves, and modest bloomers. Early girls' basketball teams played a half-court game with the forwards on one end of the court and the guards on the other end. It was believed that a full-court game was too strenuous for females.

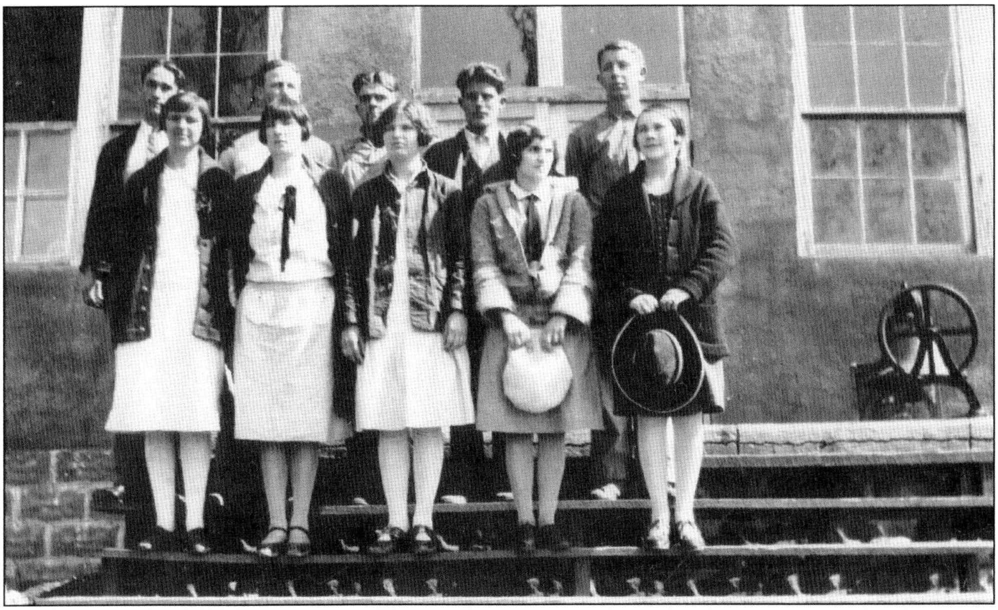

After a round of school consolidations in 1925, the county built a second public high school at Tremont, providing access to secondary education to students in eastern Itawamba County. The 1927 senior class, pictured here, elected class officers Luke Robinson, president; Neely Harbor, vice president; Sula Spearman, secretary and treasurer; Burmah Senter, poet; Retha Hiden, historian; Pauline Cofield, musician; and Louis Robinson, class fool. Note the school bell behind the students.

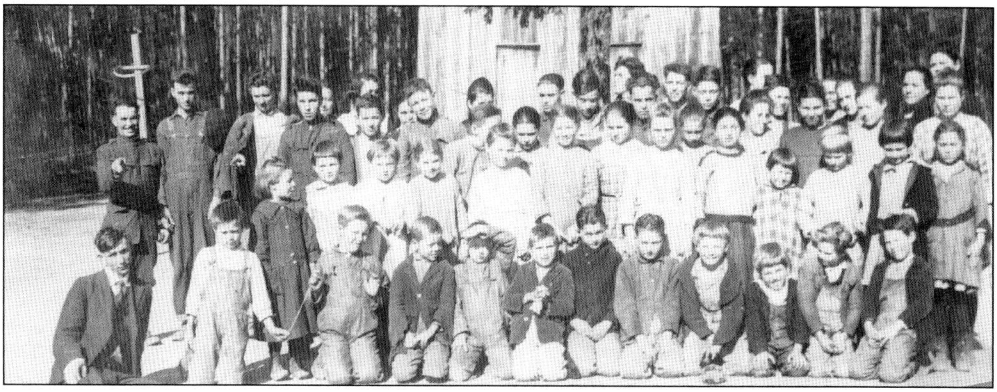

Chester A. West, seated at far left, was the teacher in 1921 at Hopewell School, located in northeastern Itawamba County. Some of the students who attended during this time were Leburn Crane, Sylvan Maxey, Lila Hartsell, Rex Maxcy, Earline Maxcy, Elmo Kent, Anna Jane Tucker, Edith Graham, and Koval Graham. Both the school and church at Hopewell are now gone, but a cemetery remains. (Courtesy of Charles and Virble Booth.)

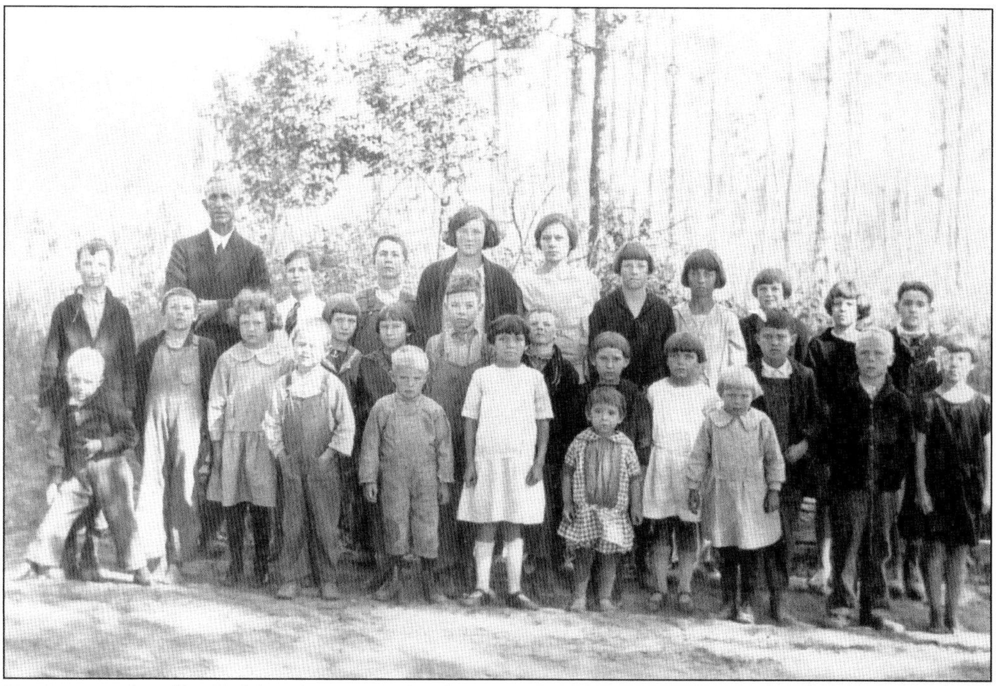

Students and staff at Sulphur Springs School in 1922–1923 are, from left to right, (first row) Burlee Shaw, Elzie Hankins, James Shaw, Vera Williams, Tarlie Spencer, and Mae Shaw; (second row) Grady Shaw, Mitchell Reed, Maynee Hankins, Mildred Gaddy, Rosalee Pierce, Jasper Gaddy, Gurley Shaw, Lafray Spencer, Era Williams, Belon Pierce, Cecil Wren, and Maloy Pierce; (third row) teacher Will Gillentine, Cecil Pierce, Audie Reed, Gracie Shaw, Winona Shaw, Lavelle Gaddy, Ruth Wren, Mazie Shaw, Violet Shaw, and Ray Pierce.

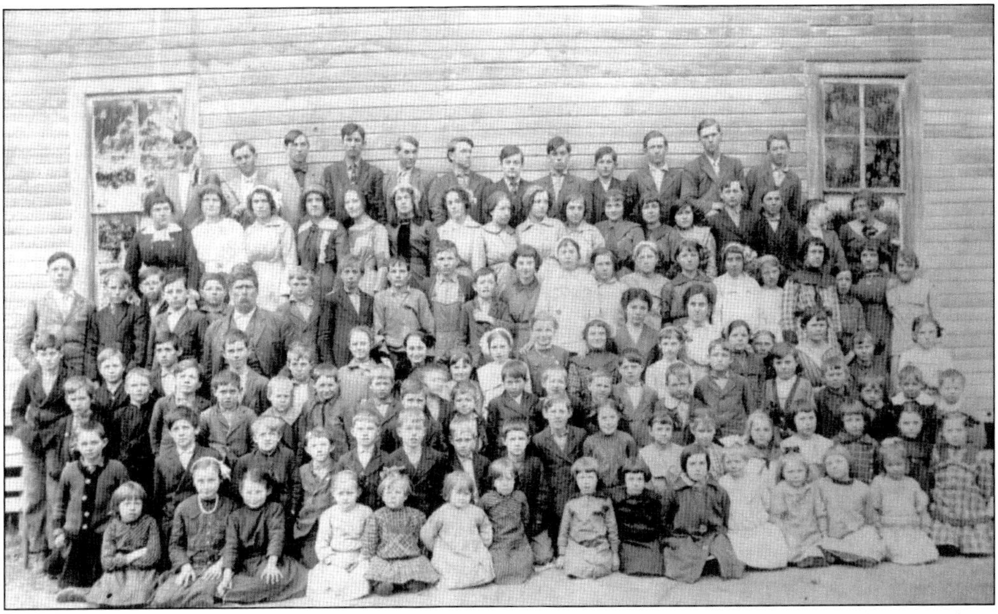

The first school serving Mantachie students was known as the Kelso School, located near today's Center Star Cemetery. In 1892, the school was moved to a new location near Mantachie Methodist Church. The new schoolhouse was a two-story, wood-frame building (shown here in 1910) that became known as Mantachie School. A 1917 newspaper reports that Mantachie School had the largest enrollment of any school in the county.

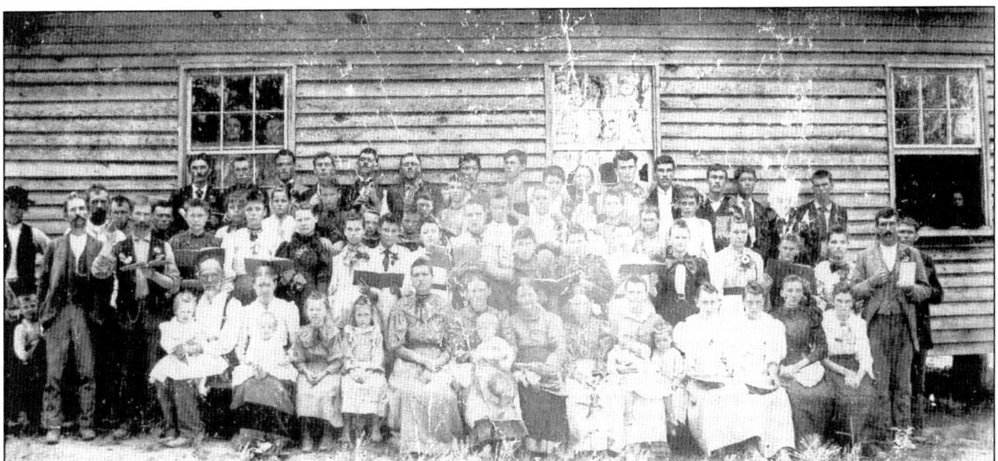

Singing schools like this one in 1896 at Antioch in northwestern Itawamba County were very popular in Itawamba County in the early part of the 1900s. These schools usually lasted two to four weeks and were conducted by local educators like Joe Sandlin or Sam Mayhall, who were musically talented, or by traveling music companies that came into the area to teach adults and children how to read music.

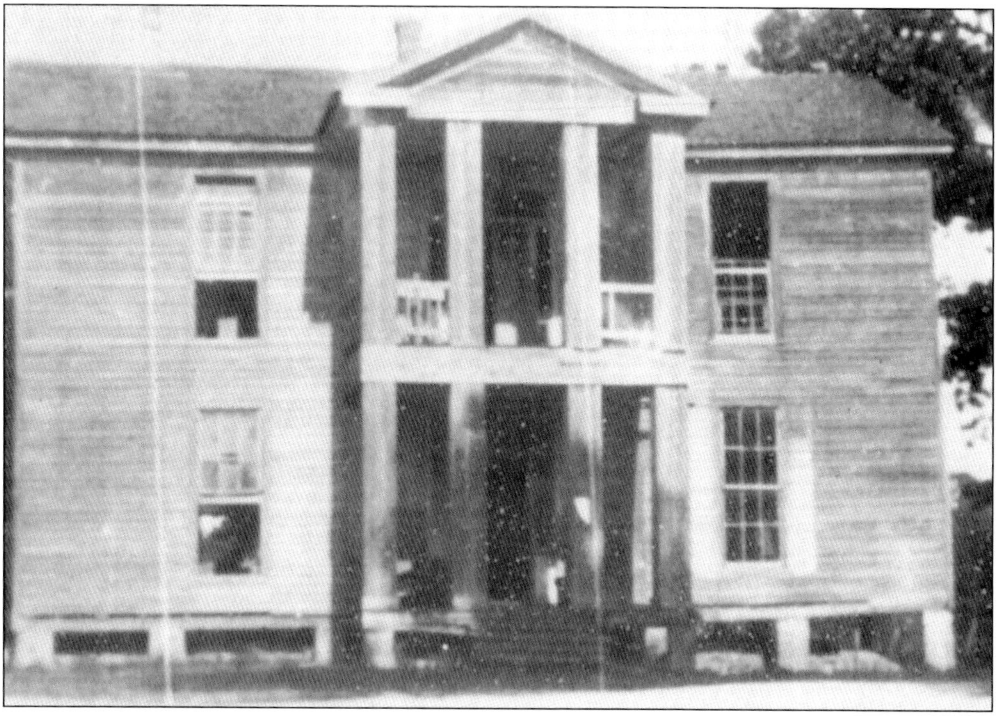

Children in Fulton attended the two-story school shown above, located on the east side of South Clifton Street, until Fulton Public School, below, was built in 1921 at the same location. The older school dated from at least the 1880s, and when no longer used as a schoolhouse, the building was purchased by G.W. Gilliland, who sawed the structure into two halves and moved them down the street for use as apartments. Fulton Public School burned in the late 1930s; following the fire, Fulton Grammar School opened about 1940 in the location on South Cummings Street. Today, the old grammar school has been renovated and is used for community events.

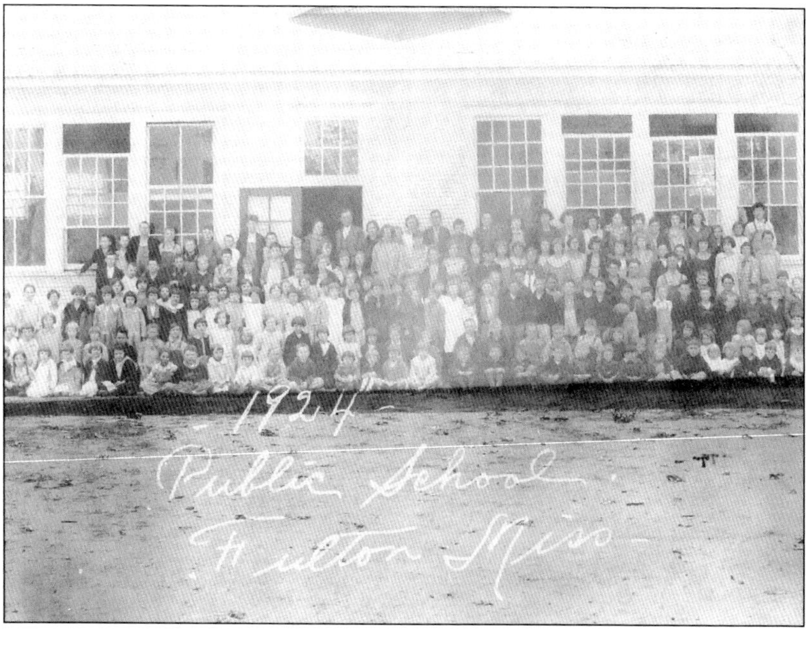

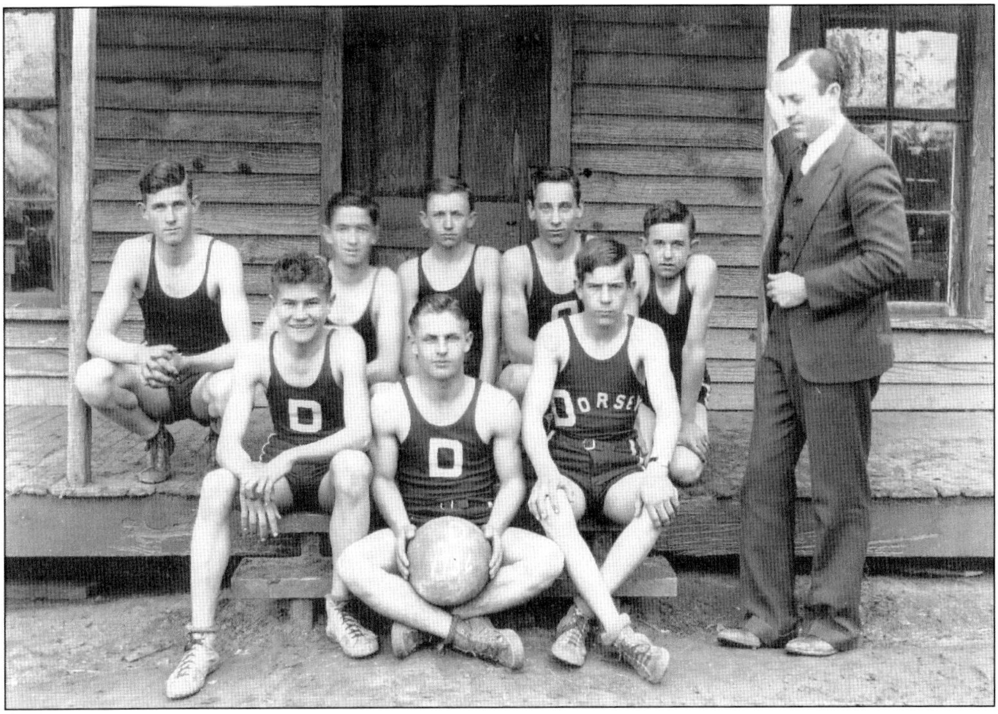

Basketball tournaments were wildly popular as schools competed for bragging rights each year at the county's championship tournament. The boys' basketball team at Dorsey School in 1936 was coached by Benson Chatham (standing, right) and consisted of, from left to right, (first row) Arrian Martin, Lavern "Buck" Gurley, and Walter Davis Sheffield; (second row) Marvin Farris, Hershell Wood, Harry York, and William Sidney Trammel. (Courtesy of Joe Nell Wood.)

Austin Red Hill School was a short-lived school, established in the early 1920s and consolidated in 1929 along with schools at Kirkville, Ozark, and Palestine into one school at Houston. Members of the 1927–1928 basketball team are, from left to right, (first row) Charley Pannell, Cullen Spigner, Bruce Spigner, and Travis Alred; (second row) Cancil Gardner, Mack Barnes, and Tracy Walker. Luther Wimbs was the coach and principal.

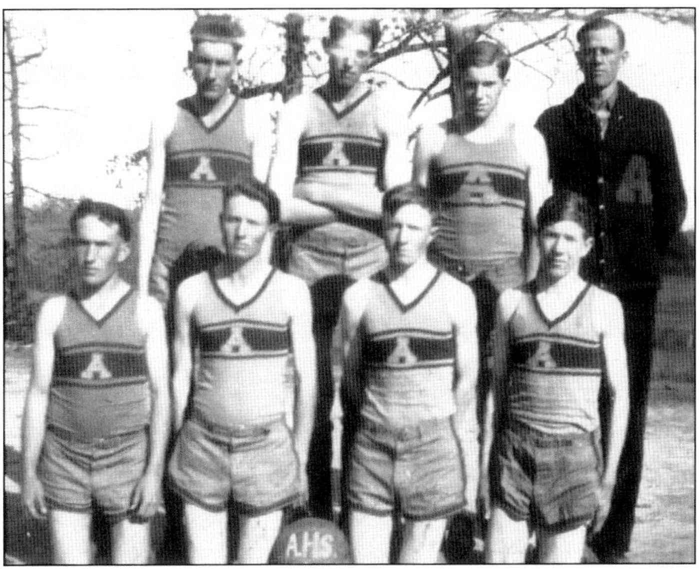

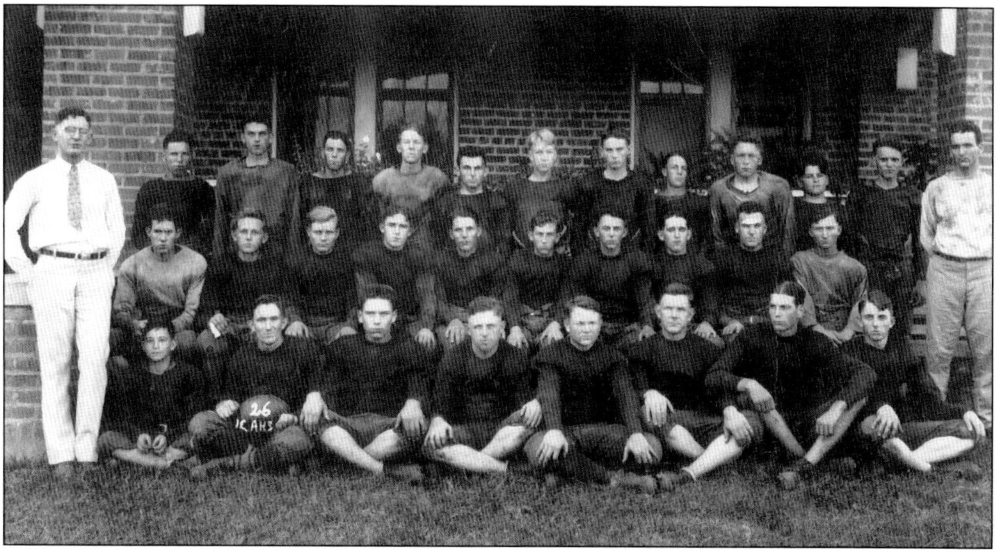

Itawamba Agricultural High School's 1926 football team consisted of 29 players. Pictured are, from left to right, (first row) Manager Oren Wren, Doyle Brown, Edgar Isabel, Tilman Collum, Burie Rutledge, Eugene Long, Erskine Thornberry, Albert Cox, John Wood Patterson, and coach Boyd Lesley; (second row) Douglas Gregory, Candler Page, Sebastian Grissom, Phillip Sheffield, Eunice Grissom, Bascum Bronson, Haskell Page, Roy Chatham, Walter Hobson, and Quitman Stovall; (third row) Arle Bean, Glenn Haynes, Oliver Love, Dillon McNeece, Mordecai Lecroy, ? Hale, Leon Bryant, ? Estes, Troy Wigginton, Stub Haynes, and L.Q. Brumley. Below, members of the Centerville basketball team in 1922 are, from left to right, Coy Chittons, Erskin Thornberry, Clinton Farrar, Milton Barber, and Ezra Grissom. Dave Raper was the team's coach. The Centerville teams were known for wearing no shoes, only socks, during their games. (Above, courtesy of Joe Rutledge.)

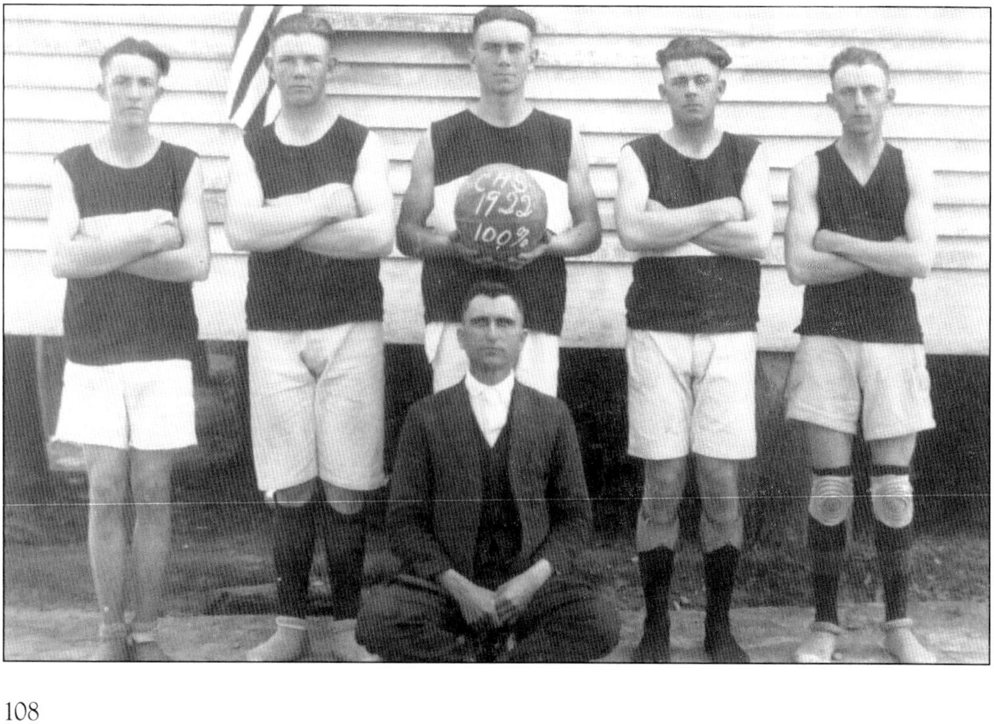

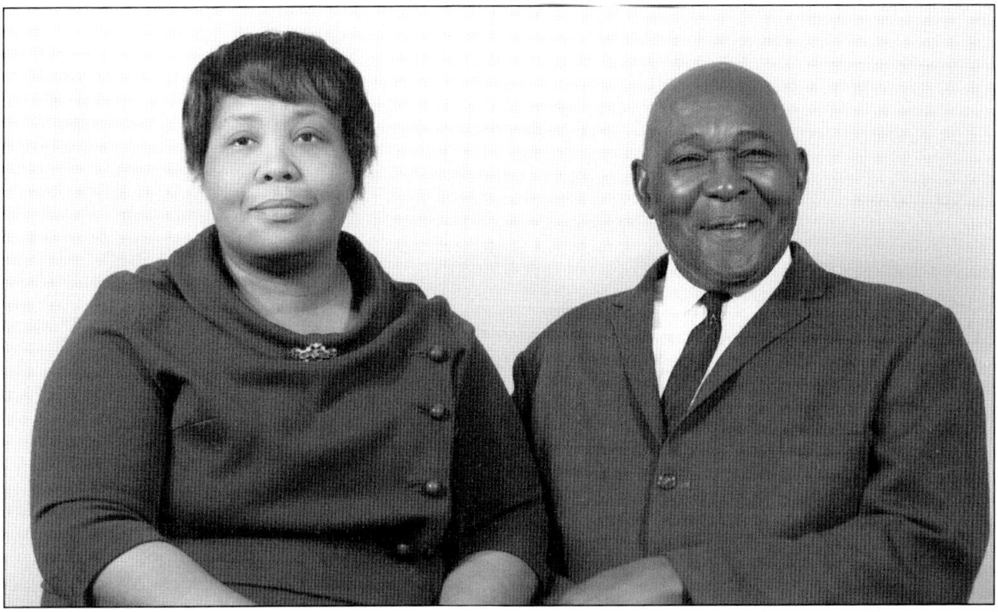

In 1941, Rev. Oliver Charles Cullins Sr. moved to Itawamba County, where he was both a respected educator and beloved pastor for nearly a half century. A graduate of Rust College, Cullins married Cecille Johnson, daughter of Seth and Laura Johnson, in 1949. Rev. Cullins served as principal for Shiloh School in Fulton and as pastor of Shiloh C.M.E. Church. (Courtesy of the Delmus Harden family.)

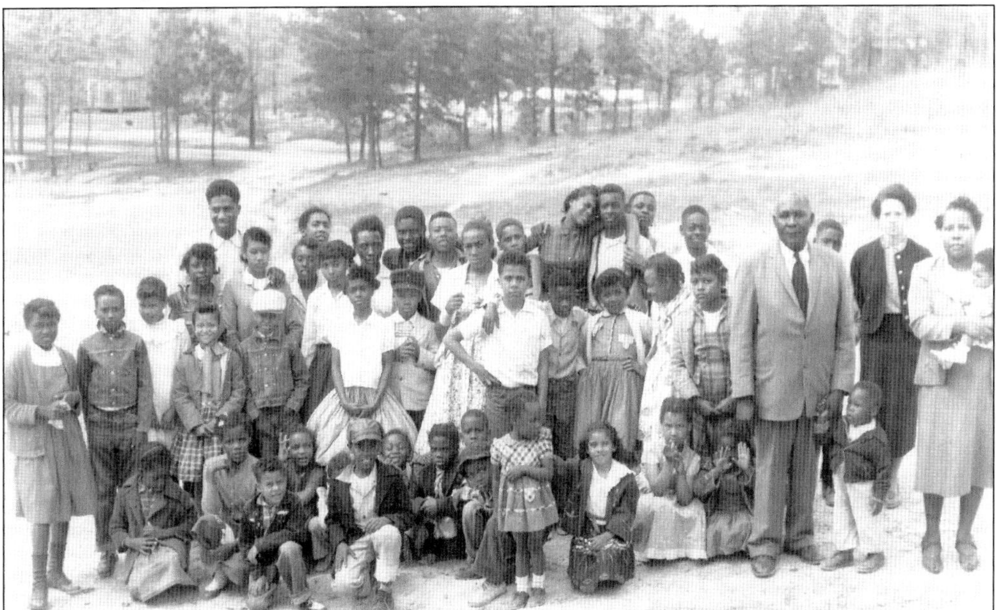

The two-room Shiloh School located in Fulton served grades one through eight. In this 1957 photograph, students are shown with their teachers, Rev. O.C. Cullins and Annie Sue Spears. Holding her son, Charles, is Cecille Cullins, wife of Rev. Cullins. Other schools serving African American students during the 1940s and 1950s were New Chapel, Pleasant Grove, Palmetto, Friendship, and Union. (Courtesy of the Delmus Harden family.)

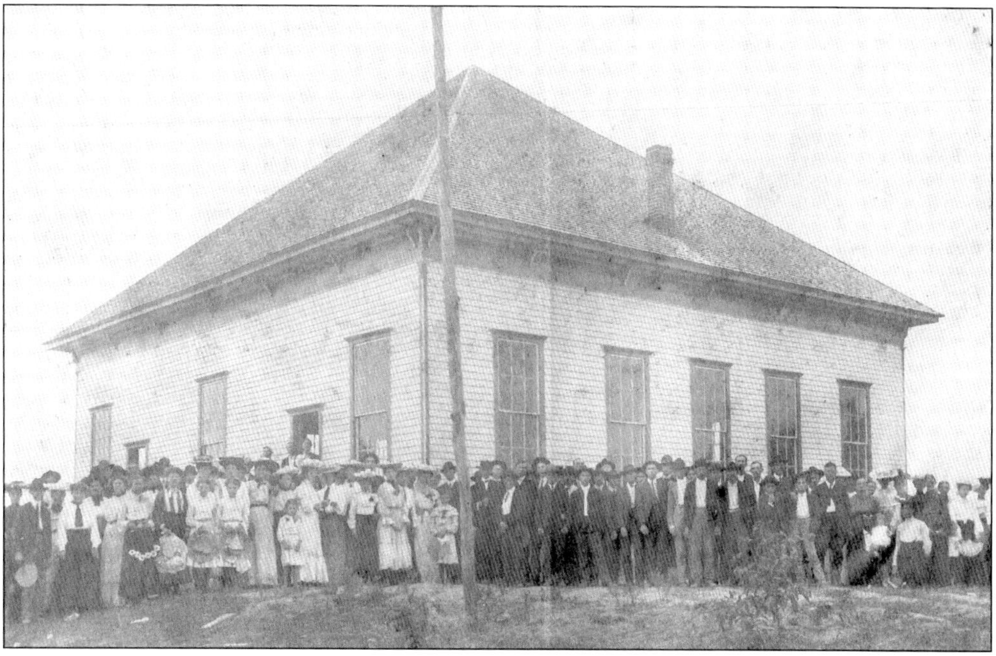

Tilden Church of Christ was organized in 1840 and had a church built near the present-day location of Union Grove Cemetery on land donated by William and Nancy Wallace. In 1905, a new church, pictured above, was built on land donated by Holly and Mandy Clayton Gray with their stipulation that no instrumental music was to be used in the services. This beautiful wood-frame church stood for over 50 years until a big storm in 1952 blew down the structure. A new church building was erected that still serves the congregation today. Below, Dow Wheeler, left, delivers a lecture at the church. The other man is unidentified.

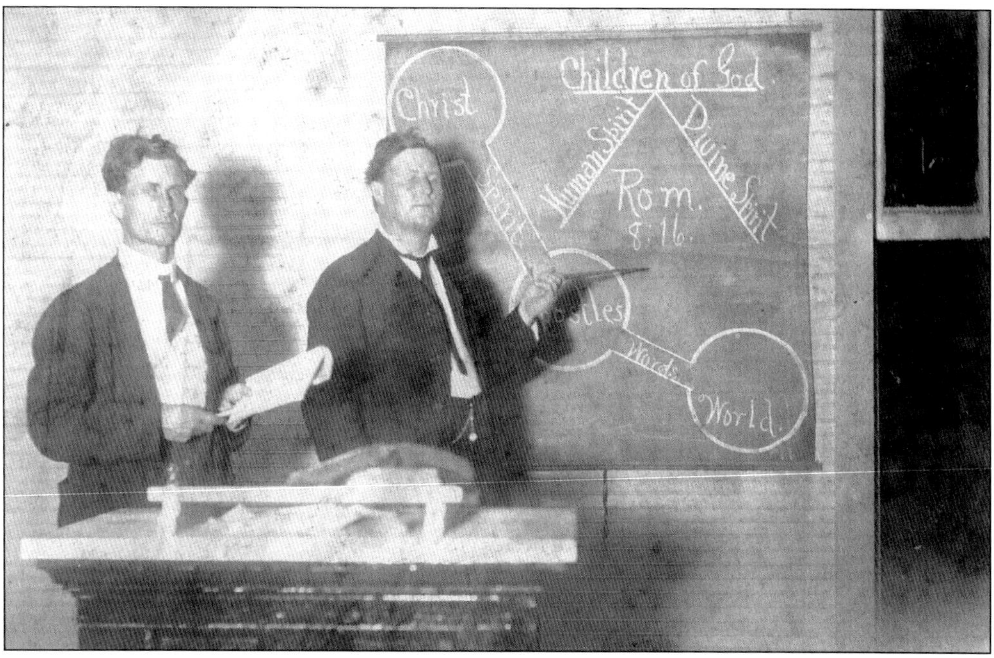

Dressed in their best suits, Baptist ministers gather for a photograph in the late 1930s or early 1940s. Those that have been identified are Ezra Gillentine, Vester Barber, J.B. "Buck" Loden, D.R. Raper, Elton Rushing, Sumpter Raburn, Greenie Gray, Jim Young, ? Marshbanks, and Martin Gray. (Courtesy of Joe Nell Wood.)

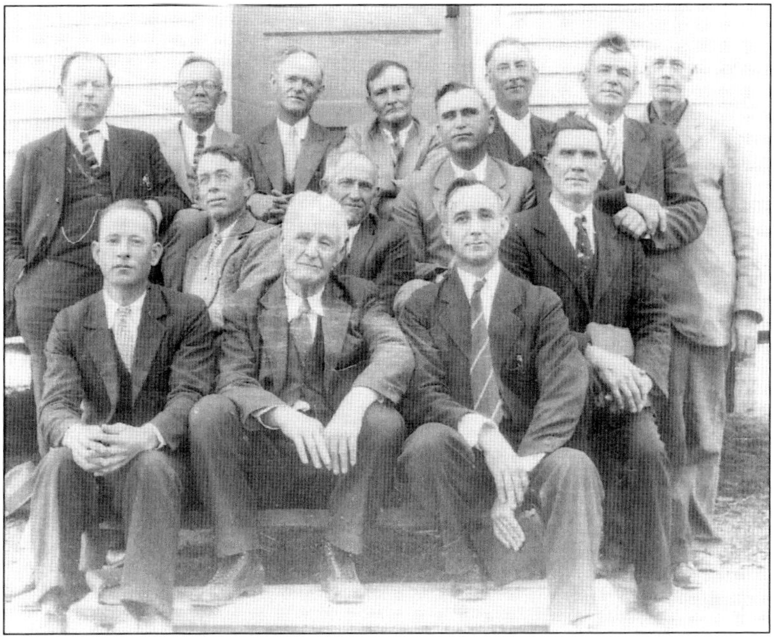

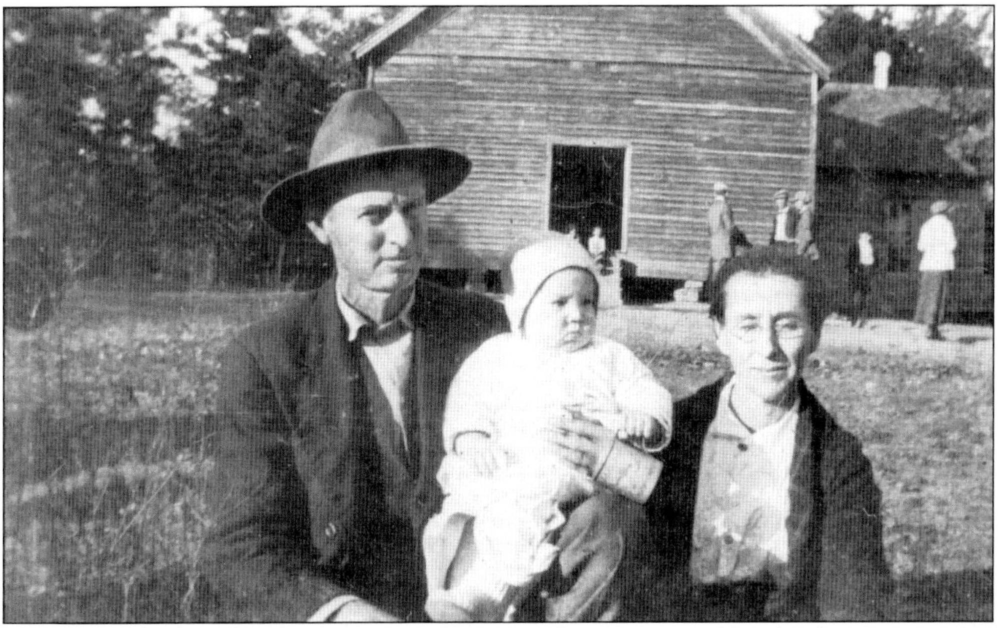

Tremont Methodist Church was organized in 1892 after a protracted brush arbor meeting that was held behind the log building of the Baptist Church. Isaac Raper served as the church's first pastor. Dow and Alice Evans Stone are pictured here in front of the original church building. In 1937, the congregation built a new church, which was bricked in 1971.

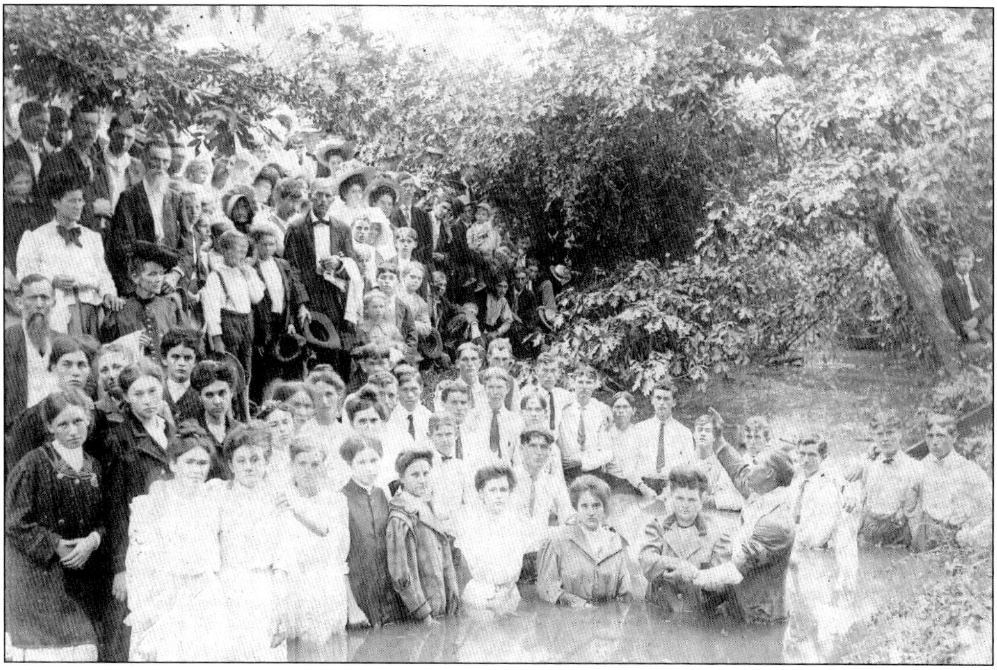

Mattox Springs, located between Dorsey and Evergreen on land owned by James P. Mattox, was a popular spot for picnics and baptisms, especially for members of Hopewell Baptist Church, whose minister, Rev. W.O. Thornley, baptizes new believers in this scene from around 1900. Bethany Hessentine "Hessie" Wood Bourland stands behind the third girl from the left in the front row. (Courtesy of Nancy Goodwin Bruce.)

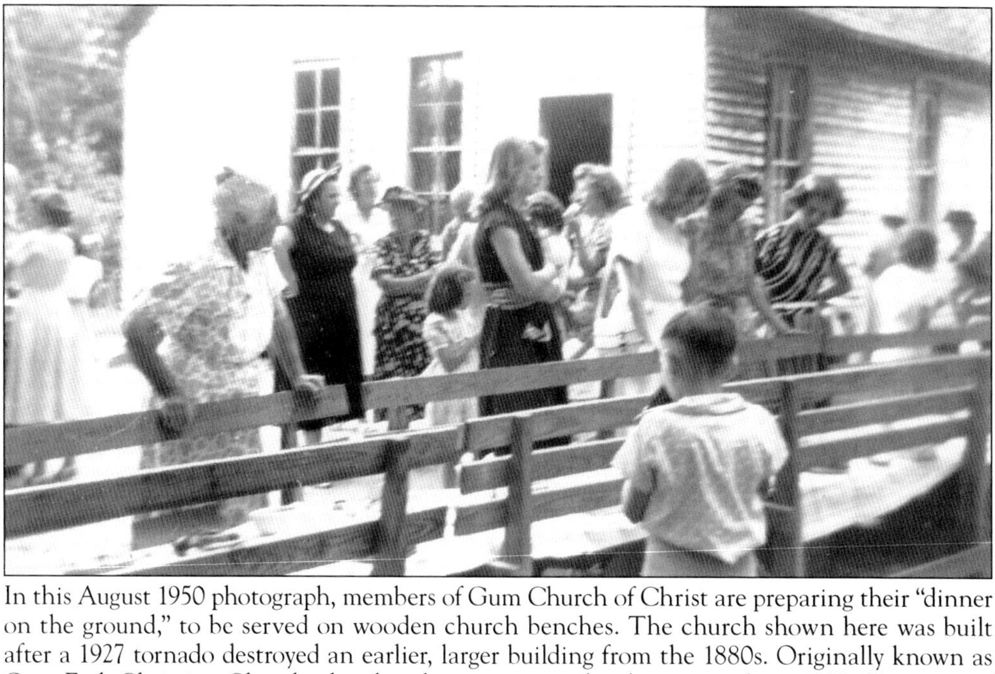

In this August 1950 photograph, members of Gum Church of Christ are preparing their "dinner on the ground," to be served on wooden church benches. The church shown here was built after a 1927 tornado destroyed an earlier, larger building from the 1880s. Originally known as Gum Fork Christian Church, the church was organized at least as early as 1839. (Courtesy of Janie Alexander.)

Dr. Hatch Cummings and B.Y. Cummings sit on the steps of the Fulton Methodist Church that was built by their uncle, Malachi C. Cummings, around 1880. This building was used until 1930, when a brick structure was built slightly east, where the Catholic church sits today. In 1961, the Methodist congregation moved two blocks east to a new building at its present location. Fulton Methodist Church was organized before 1839.

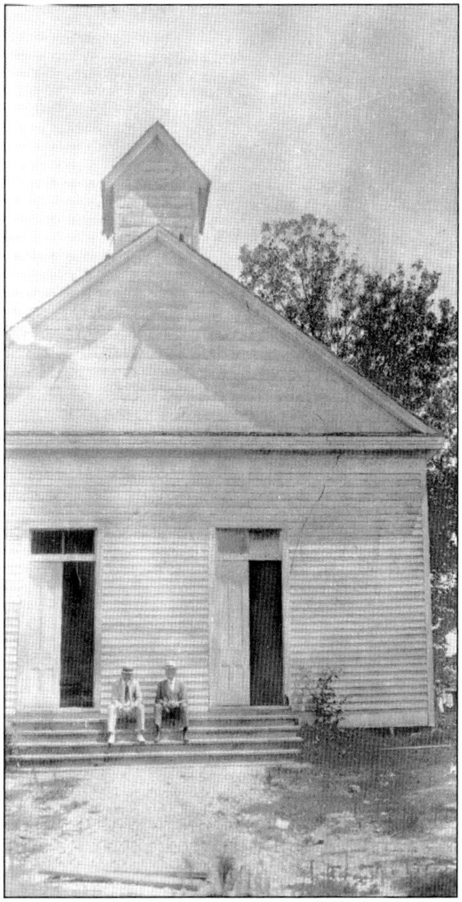

The old Pine Grove Church of Christ stands in the background as Alton Kuykendall (left) and Omie Johnson visit with one another. Omie's father, Napolian or "Polie" Johnson, was a charter member of the church, which also was known as Johnson Schoolhouse since the building served both purposes. After the original structure burned in 1927, Johnson donated land for a new building amongst the pine trees. (Courtesy of Terry Wilemon.)

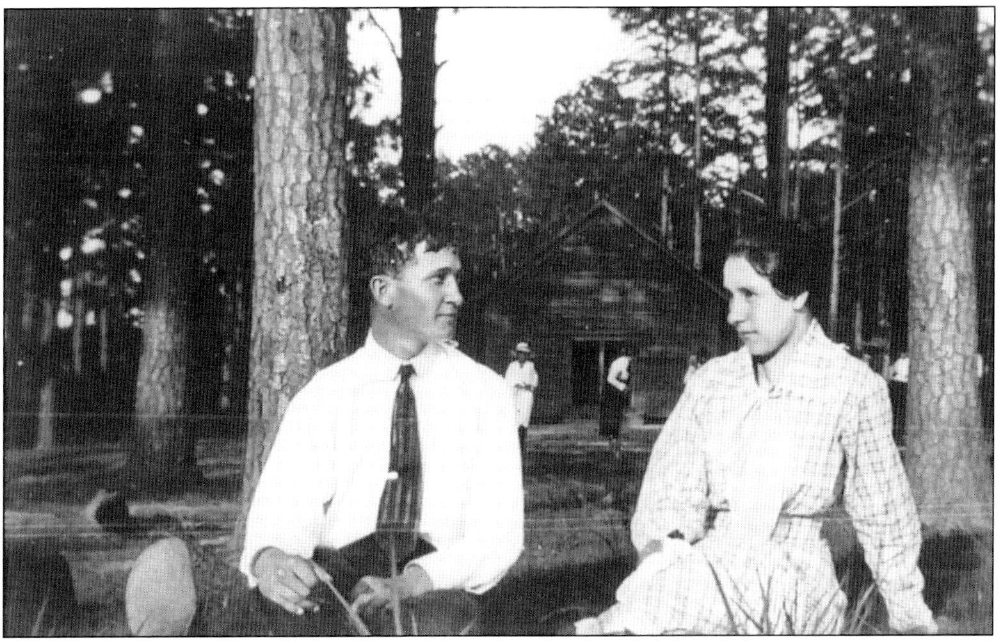

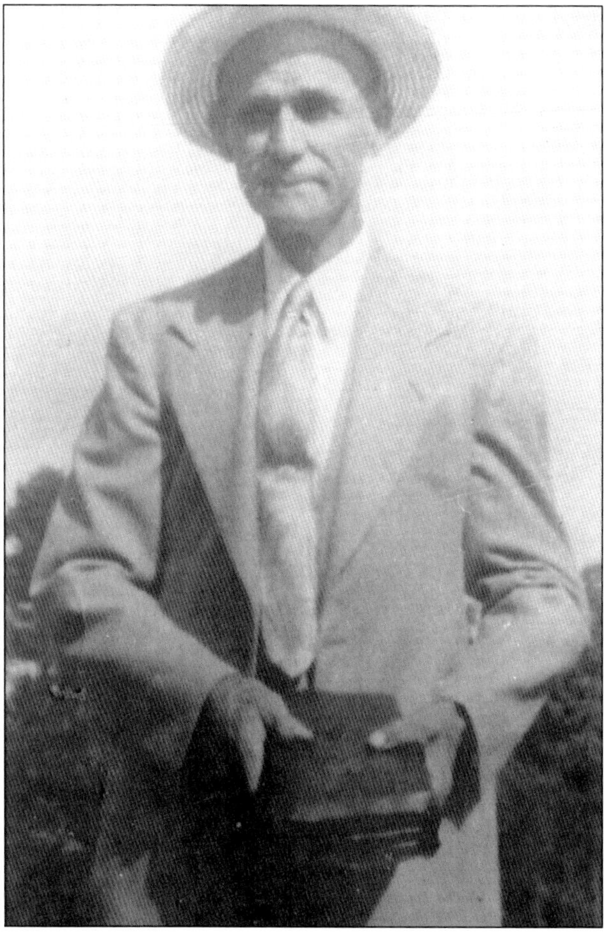

Rev. James Cicero Graham was a beloved minister of the Fairview community. He pastored Fairview Baptist Church for 37 years, having been ordained on August 4, 1922, and also served on the executive board of the Mississippi Baptist Association. Because of his musical ability, Rev. Graham taught many singing schools in the county. (Courtesy of Wilda Senter Ivy.)

Roddie Bennett's lake is the location for this baptismal scene following a week of revival services in August 1939 at Sandy Springs Baptist Church. Bro. Lunsford Thomas, Bro. Burl Franks, and Bro. ? Stone, pictured in front, were the ministers that day. Beaver Lake and the spillway from Mud Creek Mill were other commonly used locations for baptisms in the area. Bennett's lake became part of the Tombigbee Waterway in the 1980s.

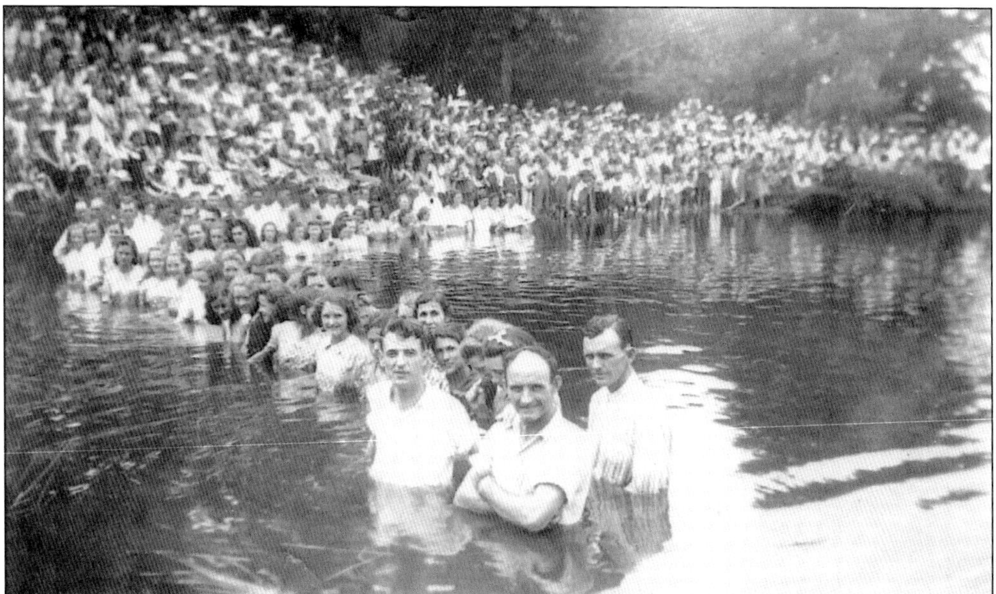

Luther Moxley was born in 1893 in Itawamba County, the son of Henry T. Moxley and Martha Arminta Sibley. He taught at several schools in the eastern part of the county until called to the ministry in 1921 while a member of Pate's Temple Church of God. In 1925, he was ordained bishop and later served as overseer for several states. Bishop Moxley was a leader within the national organization of the Church of God of Prophecy and was responsible for much of its early growth.

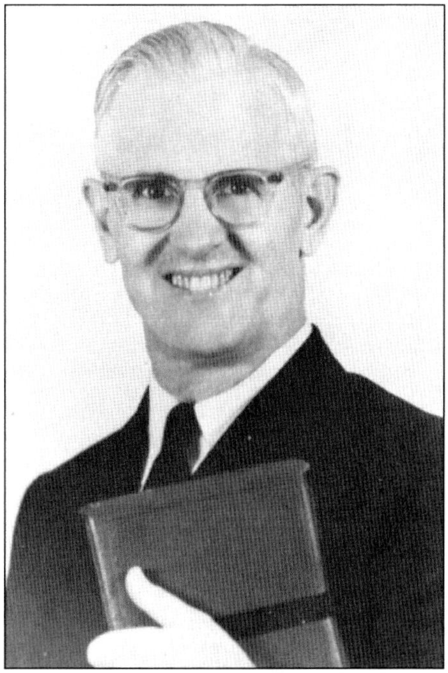

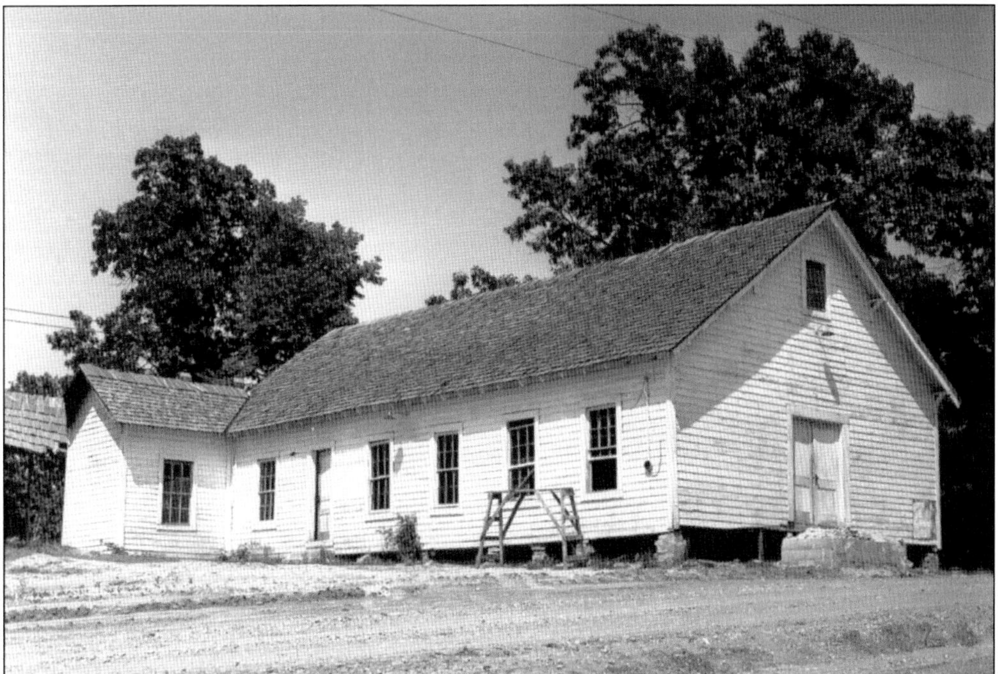

The old Sandy Springs Baptist Church building was replaced in 1964 by a brick building across the road from the wood-frame church and cemetery. Deacons of the church at the time were Lawrence Taylor, Lester Taylor, Thomas Moses, James Gentry, Troy Moses, Gordon Thomas, Trannie Jamerson, L.D. Searcy, W.S. Russell, J.B. Thomas, and R.C. Dickens. Fisher Johnson bought the old building for its lumber, and it was torn down.

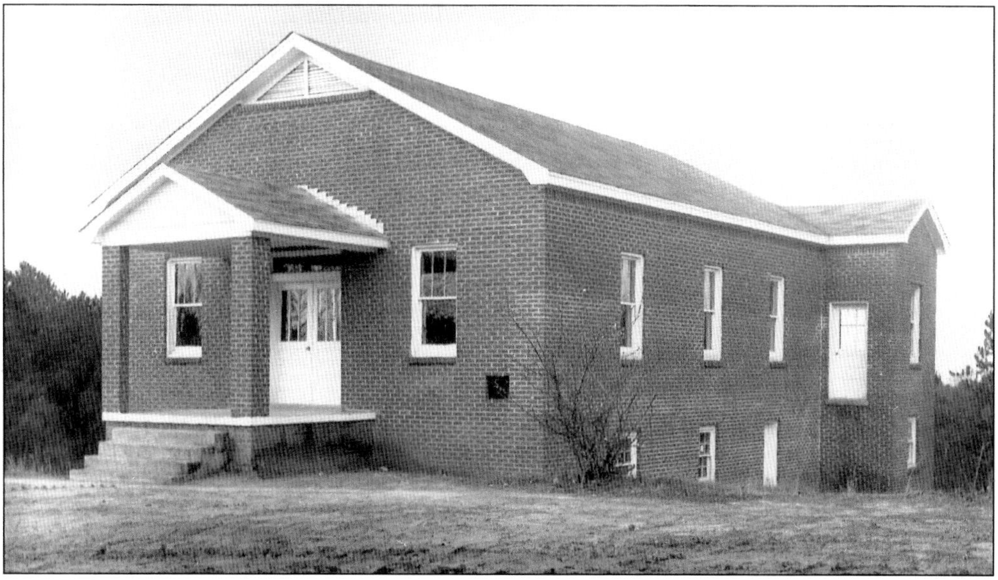

Tranquilla "Quillie" Stone and her husband, Rev. Robert Lockridge, were early organizers of St. Matthew Baptist Church, originally known as Old Berea and located east of Fulton in 1863. In the early 1890s, the church was moved to a location on the Fulton-Amory Road (Highway 25), and in 1919 another move was made to the town of Fulton, where land had been purchased in honor of Stone. Quillie Stone, Maggie Wiygul, Nan Clifton, Charlie Saltor, and Houston Rousem have been credited with much of the early success of the church, shown above in 1962. In July 1965, St. Matthew Missionary Baptist Church hosted a Vacation Bible School for its young members, shown below. Adults pictured on the far right are, from left to right, Gertrude Stone, Pastor D.D. Daniel, Nora Clifton, and Earnest Hughes. (Both, courtesy of the Delmus Harden family.)

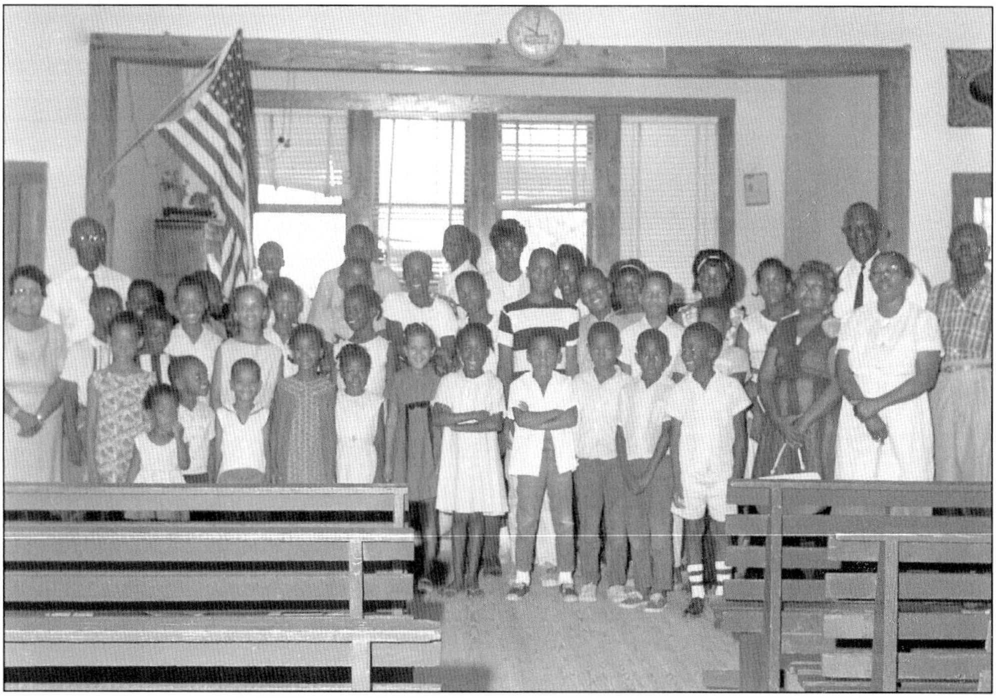

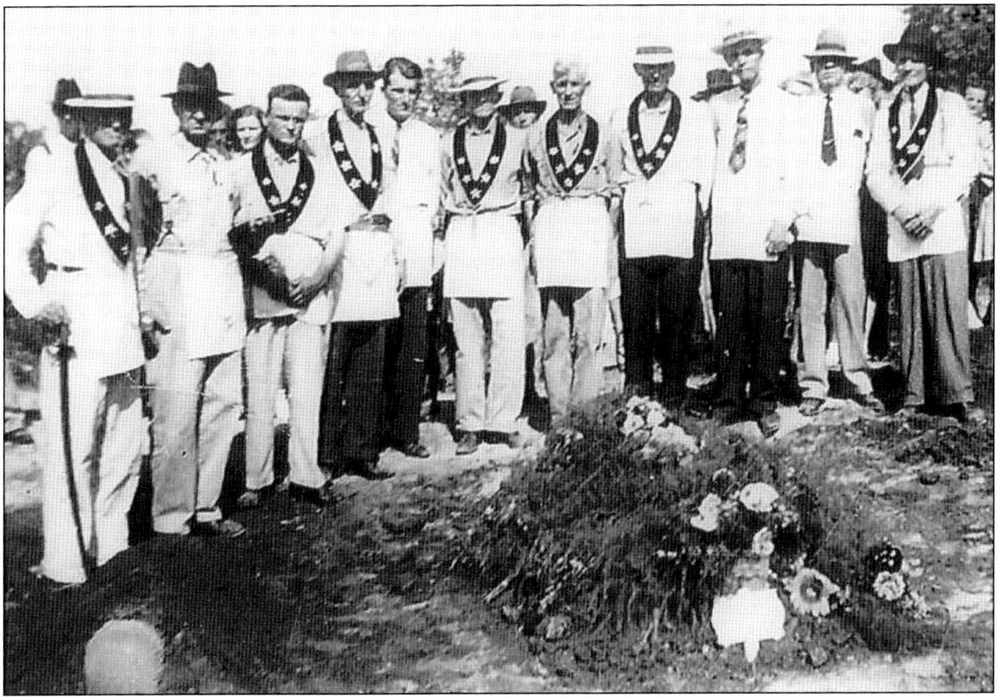

When Edgar Beam died in 1944, his funeral included a graveyard service by fellow Masons, who gathered around Edgar's grave in the cemetery at Sandy Springs Baptist Church. The men's sashes and aprons indicate a special ritual was conducted by members of the Masonic fraternity, while the evergreen branches on the grave represent immortality. Elbert Edgar Beam was the son of Samuel T. Beam and Nancy Jane Cromeans. (Courtesy of Don Dulaney.)

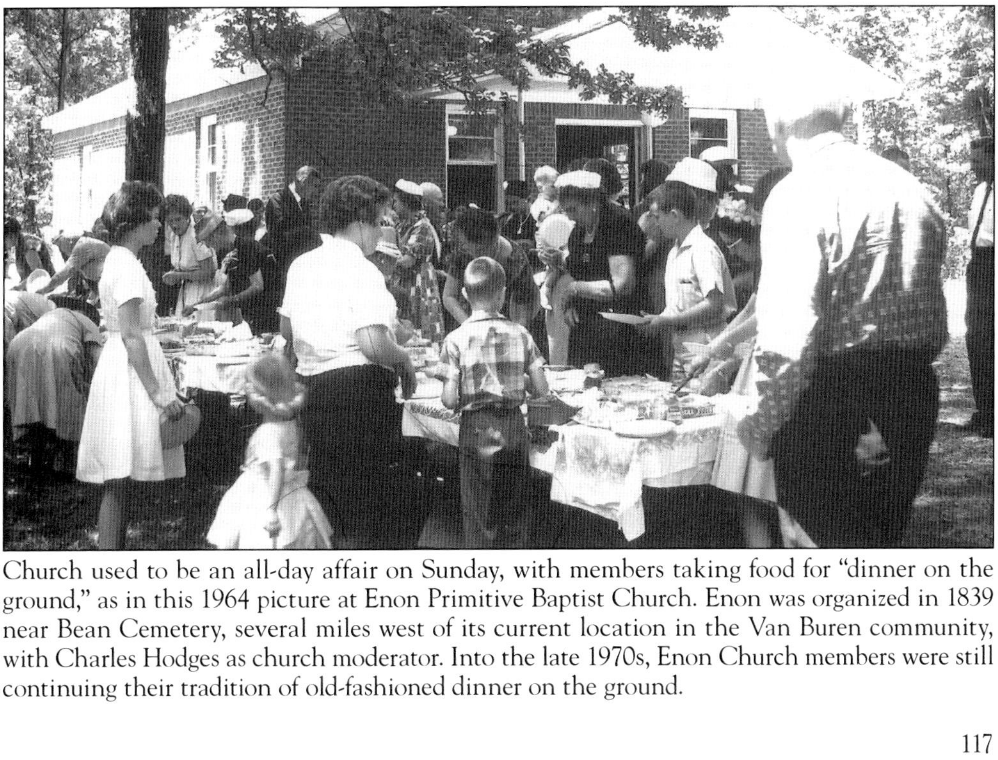

Church used to be an all-day affair on Sunday, with members taking food for "dinner on the ground," as in this 1964 picture at Enon Primitive Baptist Church. Enon was organized in 1839 near Bean Cemetery, several miles west of its current location in the Van Buren community, with Charles Hodges as church moderator. Into the late 1970s, Enon Church members were still continuing their tradition of old-fashioned dinner on the ground.

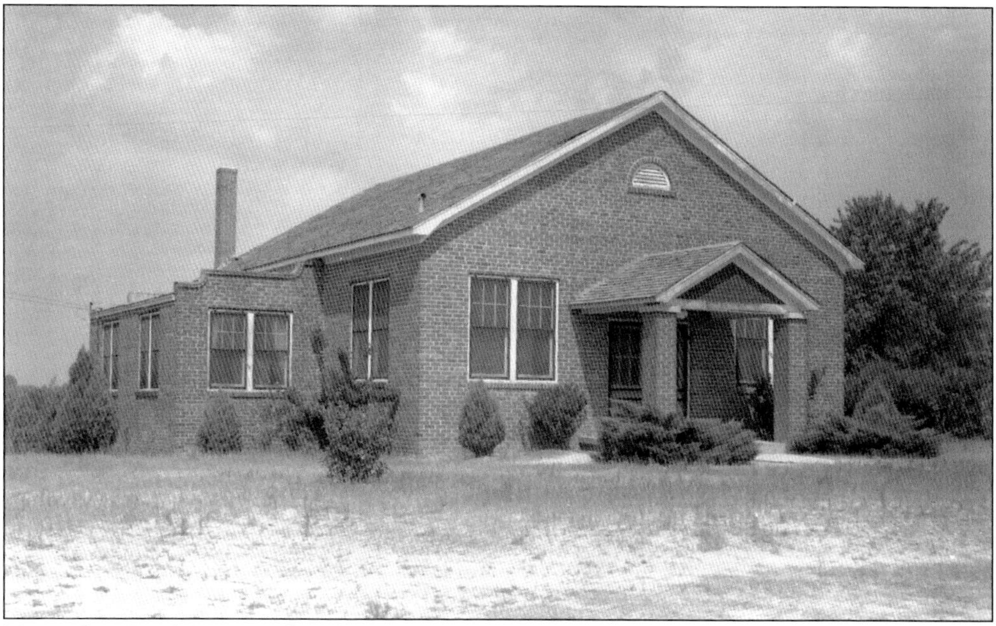

Mantachie Creek Baptist Church was organized in 1843, and a meetinghouse was erected, with William Stanfield as the first pastor. The church was later moved to a location near present-day Mantachie Cemetery. One hundred years after its founding, a new brick building, shown here in 1950, was completed just south of the former location. Today, the church is known as First Baptist Church of Mantachie. (Courtesy of the Delmus Harden family.)

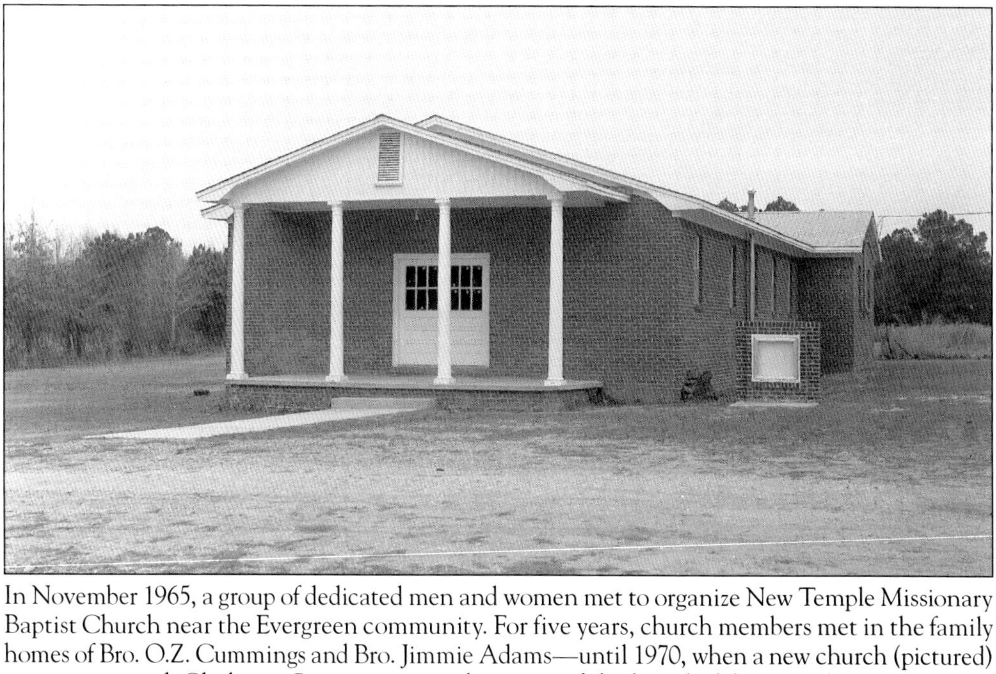

In November 1965, a group of dedicated men and women met to organize New Temple Missionary Baptist Church near the Evergreen community. For five years, church members met in the family homes of Bro. O.Z. Cummings and Bro. Jimmie Adams—until 1970, when a new church (pictured) was constructed. Chalmers Cummings was chairman of the board of deacons. (Courtesy of the Delmus Harden family.)

Seven

Sawmills and Boom Times

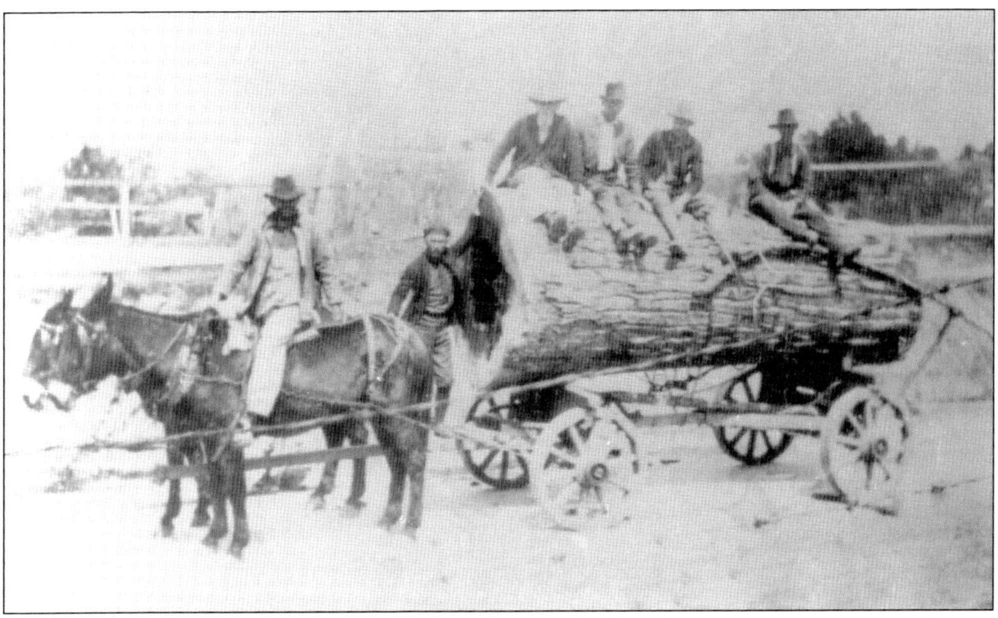

Scenes like this one were commonplace during the early 1900s, as Itawamba County's virgin forests produced lumber that was sold around the country. Massive trees were felled by the thousands during the timber booms that took place in Itawamba County. The first boom took place in the late 1890s and continued into the 1900s. After World War I, demand surged once again, and timber became a hot commodity.

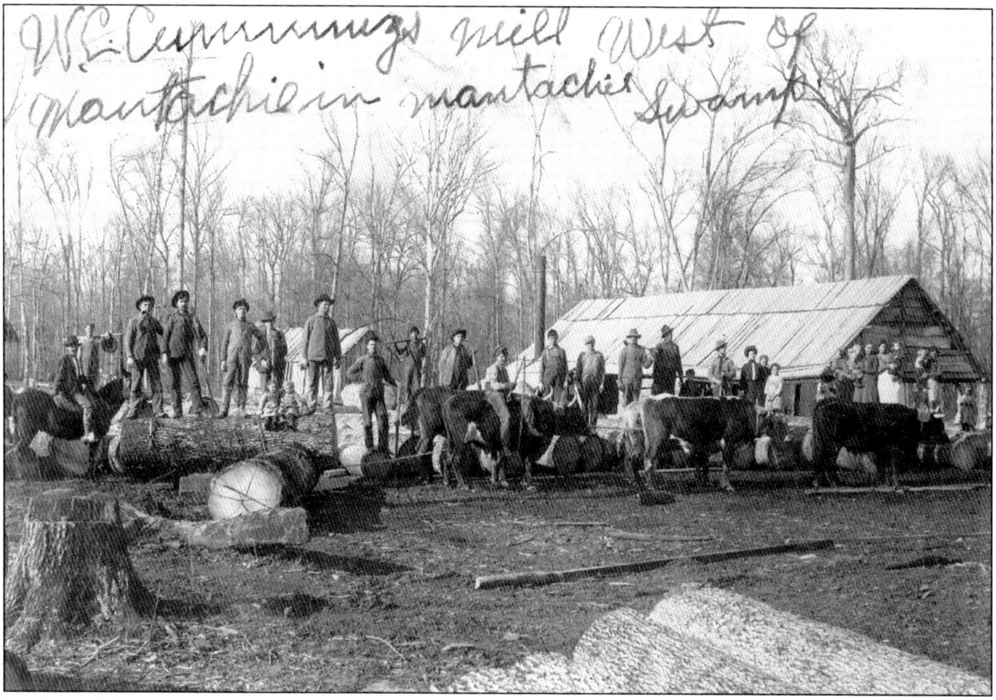

Walter Clifton Cummings Sr. owned a sawmill west of Mantachie in the early 1900s. Here, three yokes of oxen are pulling a huge log upon which stand four men and two little girls. Cummings was a timber cruiser, lumber inspector, and owner of various sawmills in Itawamba County as well as one at Waverly near West Point in Clay County. (Courtesy of Coleen Thornton, great-granddaughter of W.C. Cummings.)

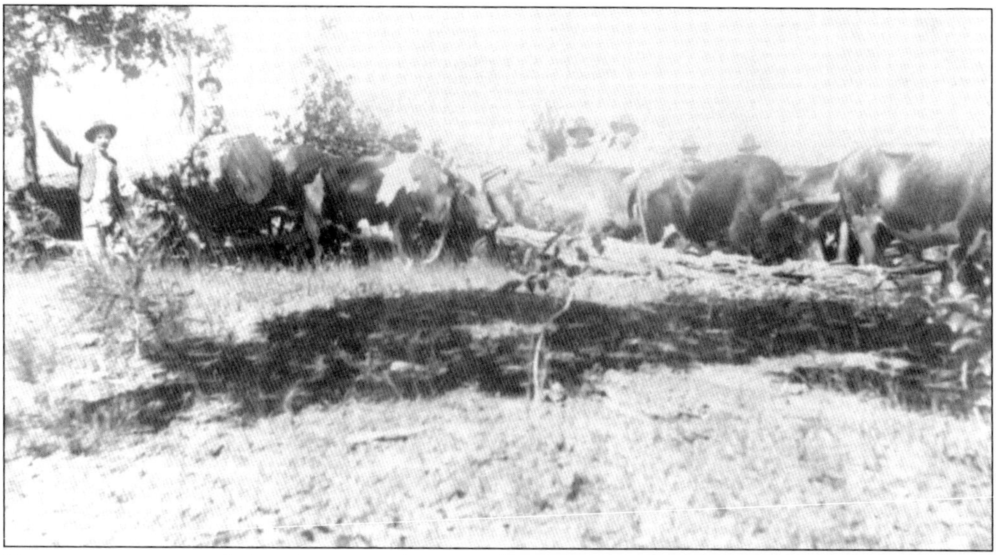

Oxen were necessary for hauling heavy logs out of river and creek bottoms, as well as remote areas without roads. The logging operation pictured took place around 1905 near Pine Forest School in the Ryan's Well community. From left to right, Henry Johnson, Eli Johnson, Rance Johnson, D. Timms, Hess Johnson, and Clovis Griffin drive oxen and logs to a water-powered mill owned by A.C. Wallace on Mud Creek.

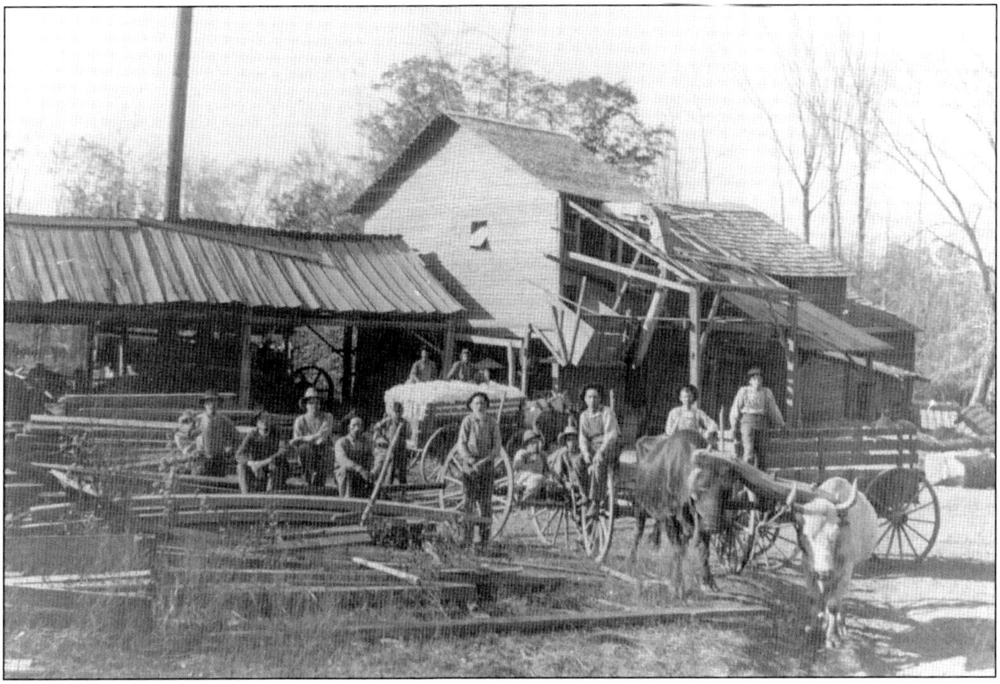

Sallis Mill, pictured here in 1908, was located between Smithville and Tremont. Owned by Elias Dejernette Sallis, it was a combination sawmill, gin, and gristmill, powered first by water, then later by steam. Pictured are, from left to right, (front) Mert Coker, Ellie Gilmore, George Cole, Lazarus Sallis, Doug Elliott, Elias Sallis, Carl Sanders, Norby Sallis, Jake Sallis, E.A. McNiece, and Bradley Sanders; (on the cotton wagon) Curtis Lockridge and Carl Elliott.

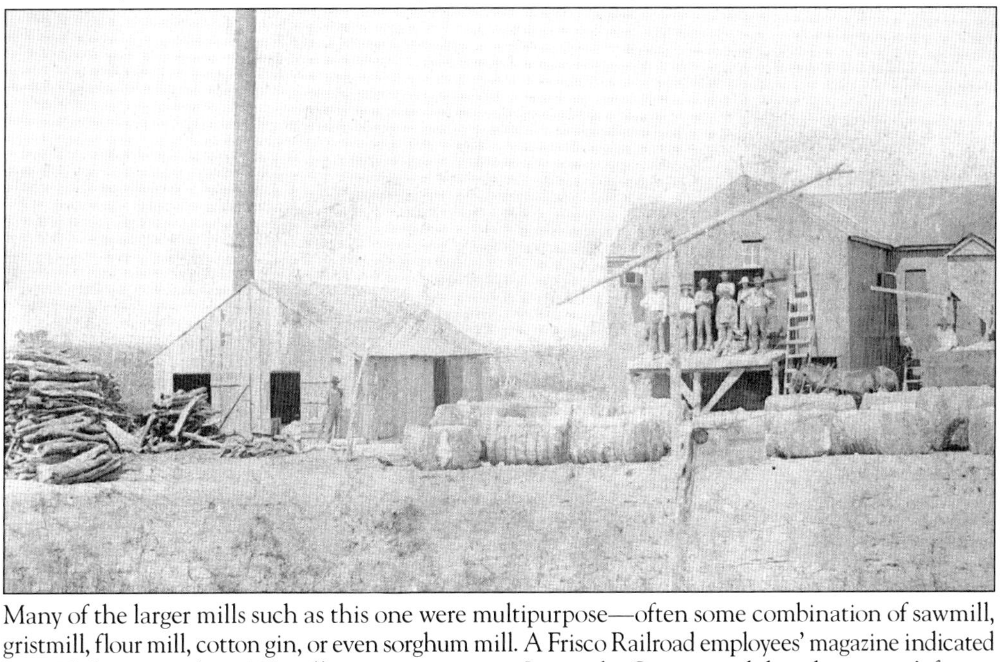

Many of the larger mills such as this one were multipurpose—often some combination of sawmill, gristmill, flour mill, cotton gin, or even sorghum mill. A Frisco Railroad employees' magazine indicated in 1925 that more than 465 mills were operating in Itawamba County and that the county's forests contained up to 400 million feet of lumber, a supply that would take 15 years to cut.

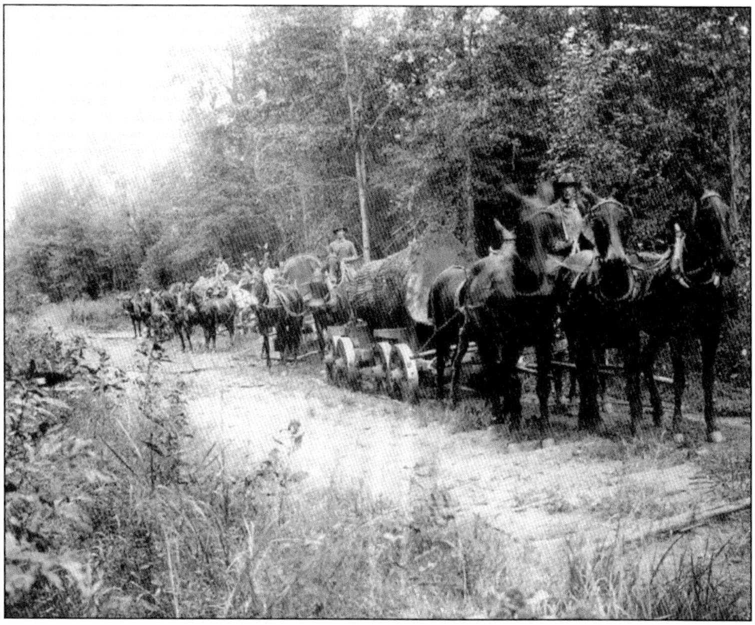

Hardwood logs are being hauled out of the forest in 1923 by at least four teams of mules pulling iron-wheeled wagons. Cut timber was typically either carried to a portable sawmill set up in the woods or to the river for floating downstream to a larger mill, or possibly transported to a specially laid logging rail line that carried the logs to a mill site. (Courtesy of Bob Franks.)

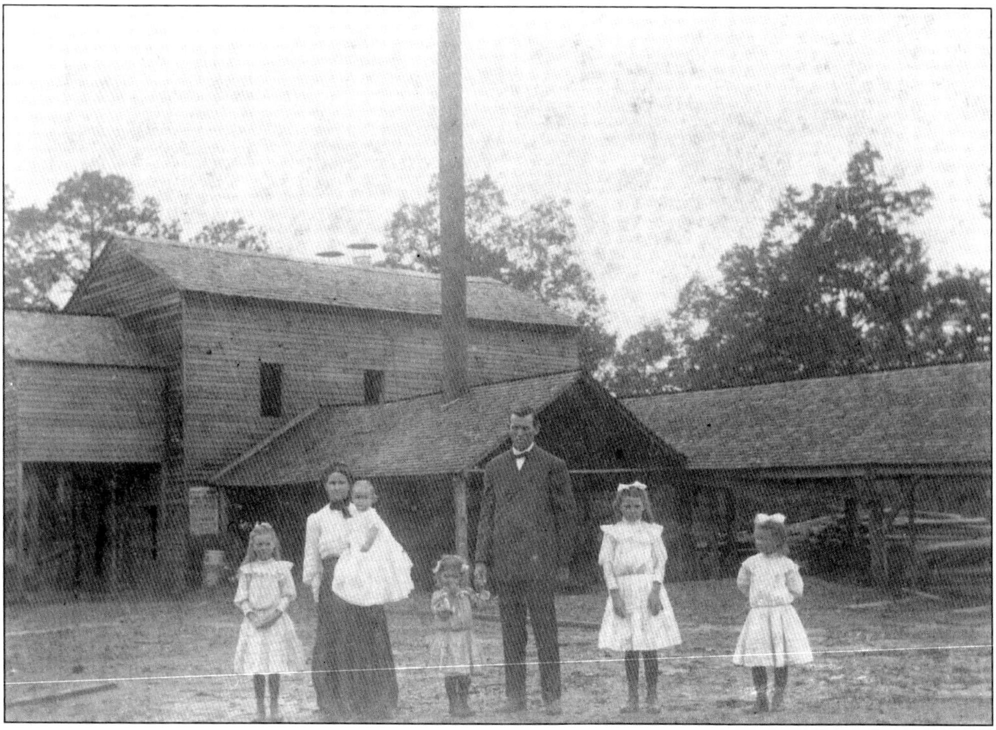

Edgar Trannie Bowen poses in 1907 with his family in front of his sawmill that was located across from the Union Grove Cemetery south of Fulton. Family members are, from left to right, Vernie, wife Narocia Stone Bowen holding daughter Maulsie, Gracie, Trannie Bowen, Verbal, and Mable. Trannie was the son of Richard "Dick" Bowen and Martha A. Webb.

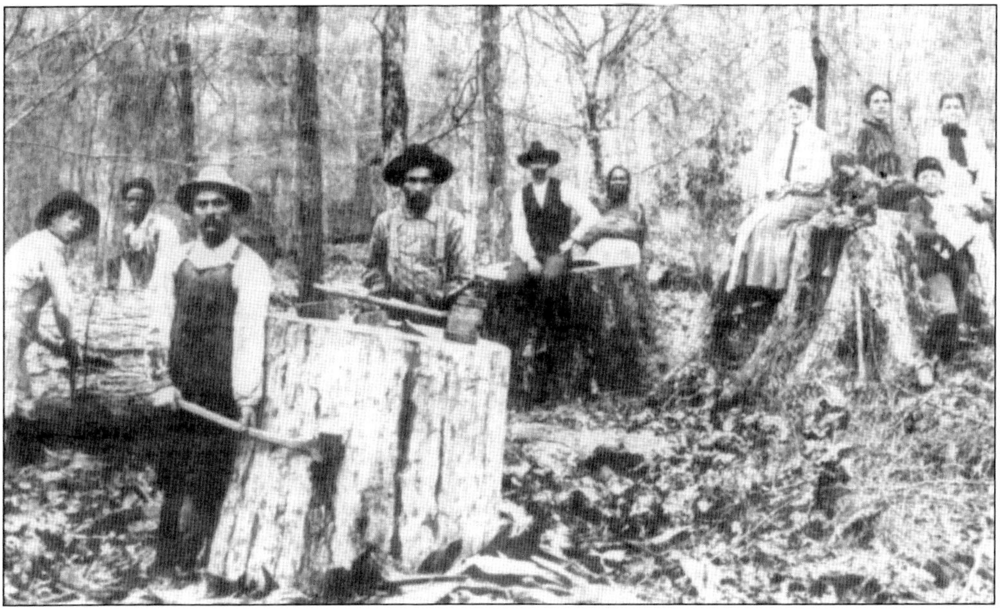

Augustus Pearce owned a mill in the Tombigbee bottom southeast of Mantachie that produced staves, strips of wood used in the making of barrels. Several stave mills existed in Itawamba County around the turn of the 20th century; some could be easily relocated to a spot of fresh timber. The mills required a steady water source for their steam boilers, which powered the saws. Boiler explosions were always a potential source of danger. Above, felled timber left huge stumps, showing the size of the trees. Logging was done by hand using two-man crosscut saws and axes. Below, workers surround a crudely built canopy that covers the sawing operation of Pearce's Stave Mill. (Both, courtesy of Bob Franks.)

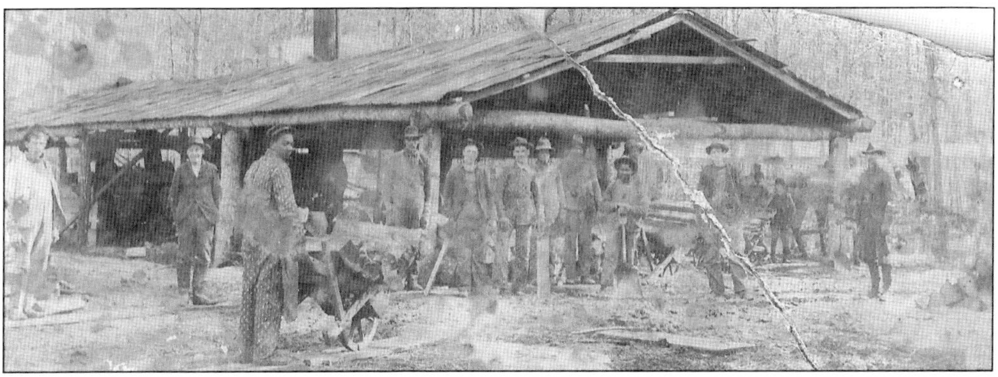

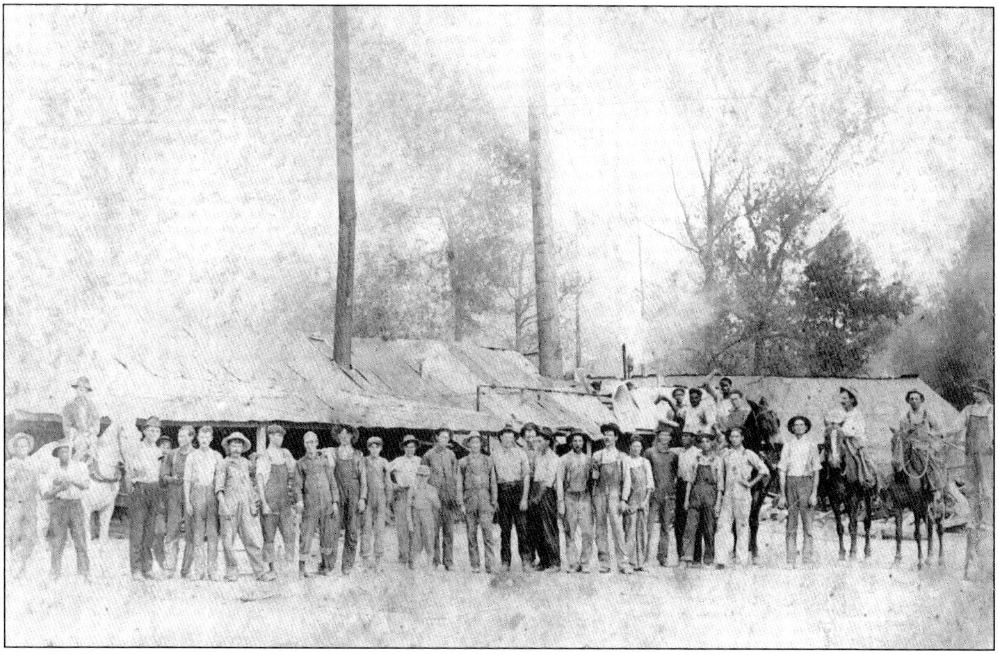

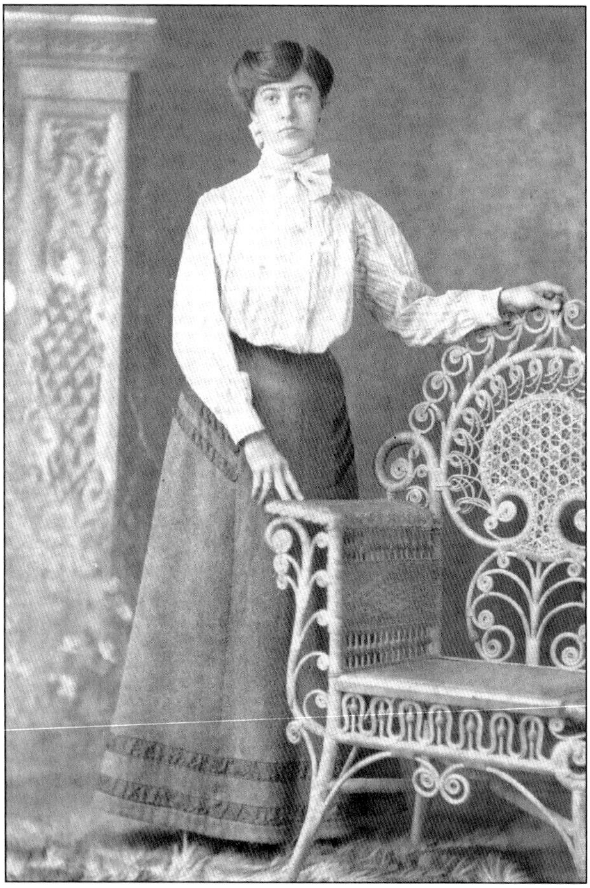

This stave mill, photographed about 1915, was located near Kirkville in western Itawamba County. Owned by Harlan Morris Manufacturing Company, the mill was a branch of the Jackson, Tennessee, company that produced staves from white oak, red oak, ash, and gum. Several of the mill's workers came to Itawamba County from the Green River valley in Kentucky, also a lumber producing region. The Henry Brown family moved from Kentucky, after a brief stay in Alabama, to Itawamba County. They worked at the Kirkville mill. Brown's stepdaughter, Minnie Bell Duvall, pictured at left, married Arch Neal, another Kentuckian who worked at the Kirkville mill. Minnie was the daughter of Mary Morris Duvall Brown. (Courtesy of Jonnie Rodgers.)

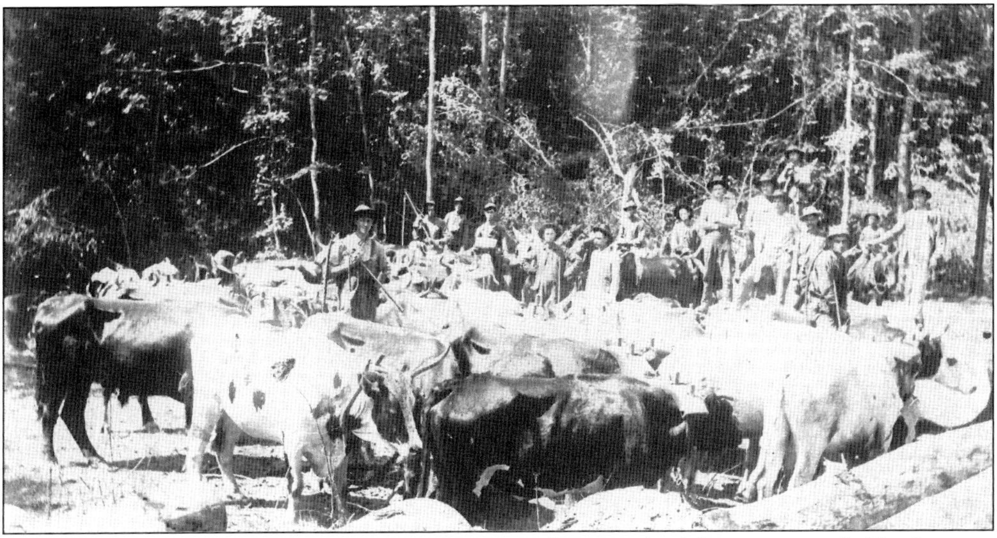

Mackey's Creek Bottom is the setting for this 1908 photograph. A logging crew, led by foreman Matt Barnes, cut timber and snaked out the logs with oxen at the headwaters of the Tombigbee River. There were four teams of oxen with four yokes to the team and two teams of mules with four mules to a team to carry the huge logs out of the bottom and to the Tombigbee River, where they were floated downstream to a mill operated by Houston Brothers Lumber Company. Some of the men in the photograph are Henry Googe, Ves Adams, Albie Perry, Fletcher Duncan, Henry Gillie, Saint Beasley, Seburn Cooley, Sherman Jamerson, and the Stacy brothers. Fisher Johnson stands at the far right.

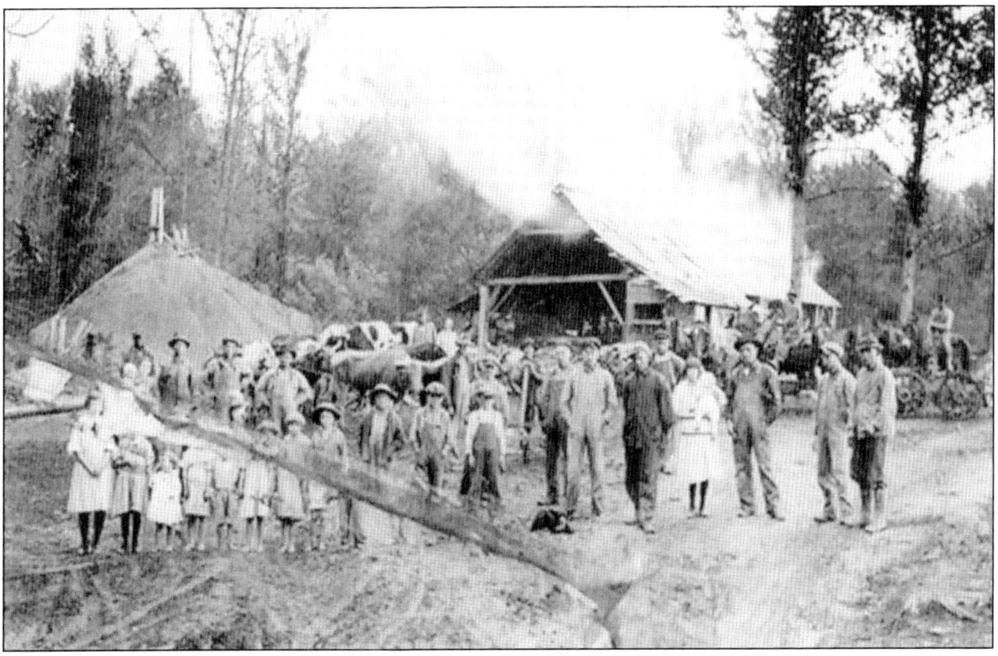

The Kentucky Lumber Company logged the creek bottoms south of Tremont. In 1920, the company purchased over 10,000 acres at $6.50 an acre and erected sawmills like this one that processed the timber and sent it to the company's planing mill at Sulligent, Alabama. This photograph probably dates from the early 1920s. (Courtesy of Rhonda Umfress Johnson.)

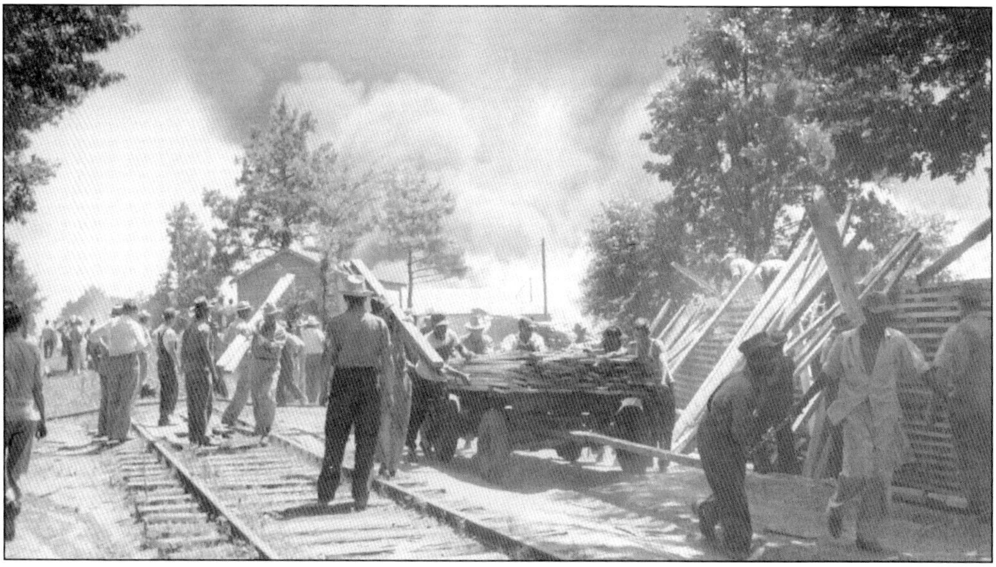

Fire was an ever-present danger at sawmills and lumber companies. E.W. Riley Lumber Mill suffered disastrous damages from a fire on June 1, 1948, at its plant in Fulton. Over two million board feet of lumber was destroyed when spontaneous combustion occurred in a sawdust pile. (Courtesy of the Delmus Harden family.)

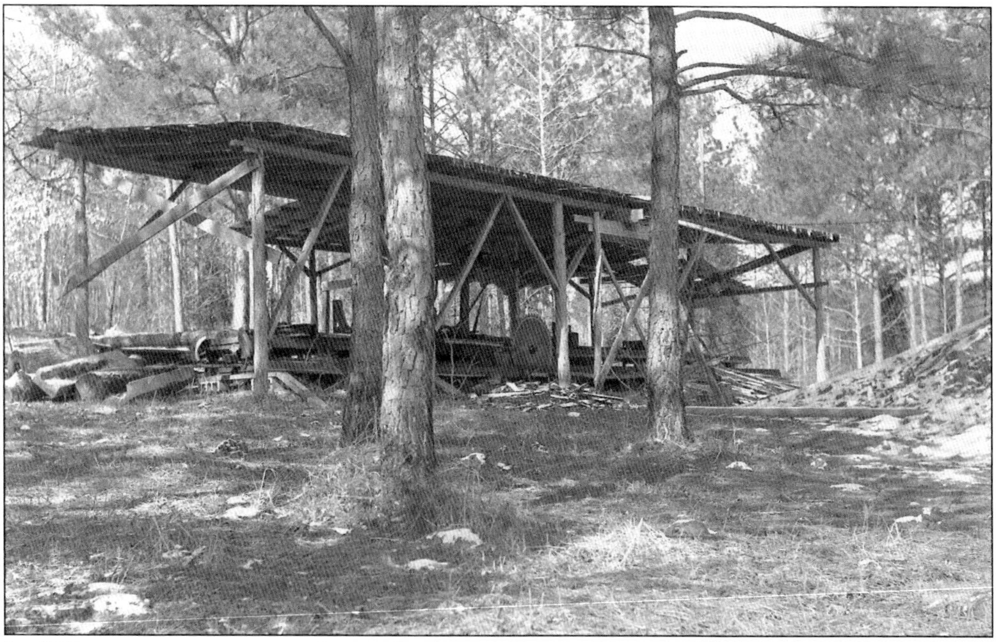

In 1970, movie crews came to Itawamba County for the filming of *Tomorrow*, a movie adapted from a 1940 short story by Mississippi's Nobel Prize winner, William Faulkner. The peckerwood mill shown here, in the Oakland community, belonged to Chester Russell and was used in scenes involving Robert Duvall's character. In the movie, the sawmill owner is referred to as "Mr. Chester Russell." (Courtesy of the Delmus Harden family.)

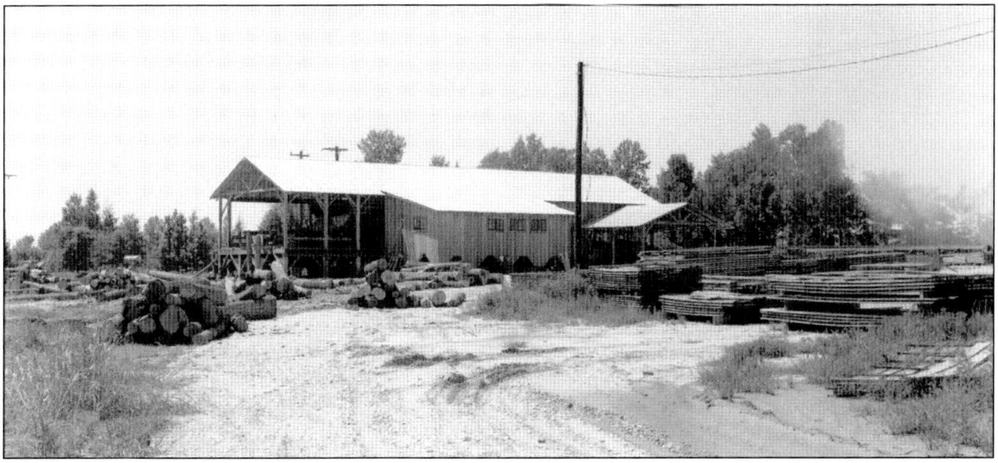

Lumber was king in Itawamba County during the 1920s. A business directory from that time lists at least 16 mills or lumber companies in Fulton alone. J.M. Gibbs Sr. came to Fulton in 1924 with the B.H. Bane Lumber Company and later started Fulton Band Mill, shown here in 1949. The Kilpatrick and Haynes families also held large lumber interests in the county. (Courtesy of the Delmus Harden family.)

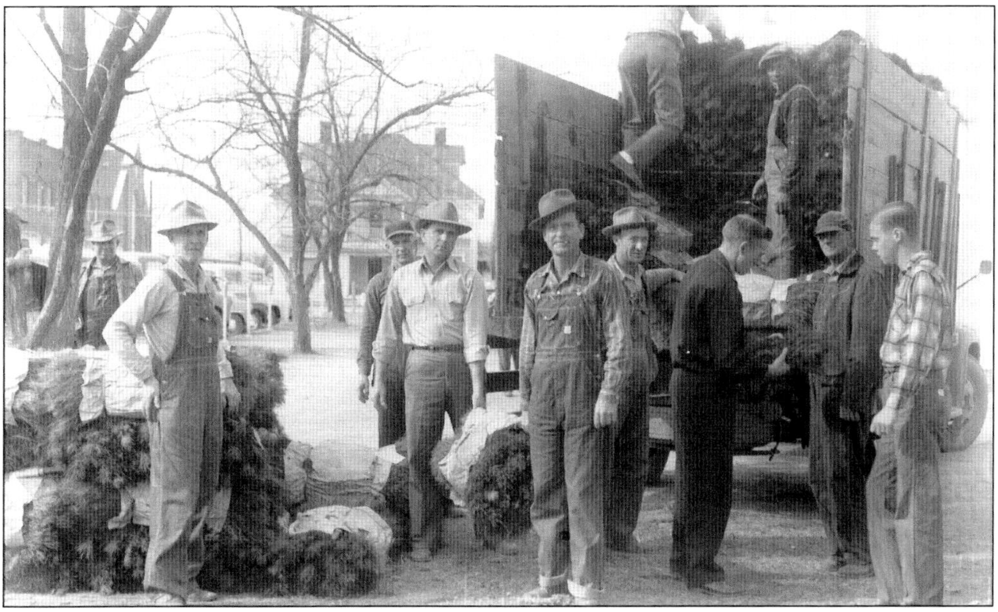

Following the logging excesses of the early 1900s, Itawambians became aware of the importance of reforestation. Thousands of seedlings were distributed each year. In this December 1955 photograph, forester Clovis Crane is on the courthouse lawn distributing bundles of pine seedlings to over 100 men. Behind the men, the former Bennett Hotel can be seen on the corner where S&W Pharmacy stands today. (Courtesy of the Delmus Harden family.)

Consistent with our mission to preserve history on a local level, this book was printed in South Carolina on American-made paper and manufactured entirely in the United States. Products carrying the accredited Forest Stewardship Council (FSC) label are printed on 100 percent FSC-certified paper.

MADE IN THE USA